CU00660430

1 MONTH OF
FREE
READING

at

www.ForgottenBooks.com

By purchasing this book you are eligible for one month membership to ForgottenBooks.com, giving you unlimited access to our entire collection of over 700,000 titles via our web site and mobile apps.

To claim your free month visit:

www.forgottenbooks.com/free503382

* Offer is valid for 45 days from date of purchase. Terms and conditions apply.

ISBN 978-0-483-78677-6
PIBN 10503382

This book is a reproduction of an important historical work. Forgotten Books uses
state-of-the-art technology to digitally reconstruct the work, preserving the original format
whilst repairing imperfections present in the aged copy. In rare cases, an imperfection in
the original, such as a blemish or missing page, may be replicated in our edition. We do,
however, repair the vast majority of imperfections successfully; any imperfections that
remain are intentionally left to preserve the state of such historical works.

Forgotten Books is a registered trademark of FB &c Ltd.
Copyright © 2017 FB &c Ltd.
FB &c Ltd, Dalton House, 60 Windsor Avenue, London, SW19 2RR.
Company number 08720141. Registered in England and Wales.

For support please visit www.forgottenbooks.com

LIBRARY

OF THE

UNIVERSITY OF CALIFORNIA.

GIFT OF

Received _____ Feb _____, 188 9 .

Accessions No. 3833 ⚡ Shelf No. _____

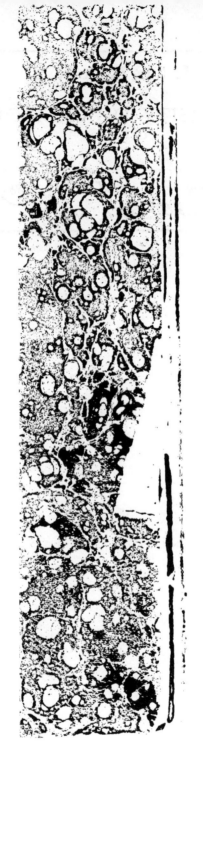

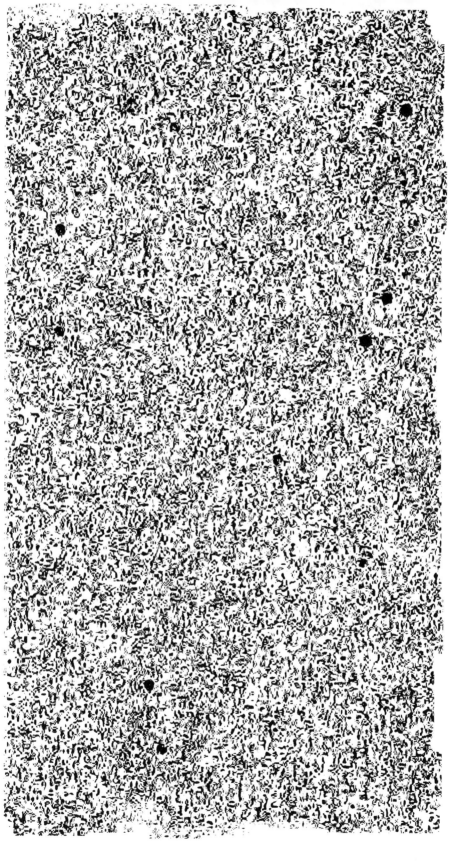

THE LIMITATIONS

OF THE

EXPANSION·OF STEAM:

FORMING THE SIXTEENTH CHAPTER OF

THE RELATIVE PROPORTIONS OF THE STEAM-ENGINE.

BY

WILLIAM DENNIS MARKS,

WHITNEY PROFESSOR OF DYNAMICAL ENGINEERING IN THE UNIVERSITY OF PENNSYLVANIA.

WITH NUMEROUS DIAGRAMS.

FROM THIRD EDITION,

REVISED AND ENLARGED.

PHILADELPHIA:

PRESS OF J. B. LIPPINCOTT COMPANY.

1887.

THE LIMITATIONS

OF THE

EXPANSION OF STEAM:

FORMING THE SIXTEENTH CHAPTER OF

THE RELATIVE PROPORTIONS OF THE STEAM-ENGINE.

BY

WILLIAM DENNIS MARKS,

WHITNEY PROFESSOR OF DYNAMICAL ENGINEERING IN THE UNIVERSITY
OF PENNSYLVANIA.

WITH NUMEROUS DIAGRAMS.

FROM THIRD EDITION,

REVISED AND ENLARGED.

PHILADELPHIA:

PRESS OF J. B. LIPPINCOTT COMPANY.

1887.

Copyright, 1887, by J. B. Lippincott Company.

PREFACE.

THE condensation of steam by the walls of the steam-cylinder is a fact whose existence has been repeatedly proved by many distinguished experimenters.

Probably Watt and a long line of followers knew that great expansions meant greater proportional losses through condensation.

We take the following paragraphs regarding Watt's work from Galloway's *History of the Steam-Engine*, page 30 (Edition of 1828).

"But in the year 1763–64, having occasion to repair a model of Newcomen's engine belonging to the Natural Philosophy Class of the University, his mind was again directed to the subject. At this period his knowledge was principally derived from Desaguliers, and partly from Belidor. He set about repairing the model *as a mere mechanician*, and when that was done and set to work, he was surprised that its boiler was not supplied with steam, though apparently quite large enough (the cylinder of the model being two inches in diameter and six inches stroke, and the boiler about nine inches in diameter); by blowing the fire it was made to take a few strokes, but required an enormous quantity of injection water, though it was very lightly loaded by the column of water in the pump. It soon occurred to him that this was caused by the little cylinder exposing a greater surface to condense the steam than the cylinders of larger engines did in proportion to their respective contents, and it was found that by shortening the column of water the boiler

3

could supply the cylinder with steam and the engine would work regularly with a moderate quantity of injection. It now appeared that the cylinders being of brass would conduct heat much better than the cast-iron cylinders of larger engines (which were generally lined with a strong crust), and that considerable advantage could be gained by making the cylinders of some substance that would receive and give out heat the slowest. A small cylinder of six inches diameter and twelve inches stroke was constructed of wood previously soaked in linseed oil and baked to dryness. Some experiments were made with it, but it was found that cylinders of wood were not at all likely to prove durable, and that the steam which was condensed in filling it still exceeded the proportion of that which was required in engines of larger dimensions. It was also ascertained that unless the temperature of the cylinder itself were reduced as low as that of the vacuum, it would produce vapor of a temperature sufficient to resist part of the pressure of the atmosphere.

"All attempts, therefore, to reduce, by a better exhaustion by throwing in a greater quantity of injection water, was a waste of steam, for the larger quantities of injection cooled the cylinder so much as to require quantities of steam to heat it again, out of proportion to the power gained by having made a more perfect vacuum."

Thus we see that Watt recognized the condensation due to the relative surface exposed, as also that due to the difference of temperature between exhaust and initial steam. With the rude machines then in use, it was out of the question for him to attempt to reduce the condensation by high rotative speeds.

In addition to the experiments of Watt, we have a large number by D. K. Clark, G. A. Hirn, and Chief Engineers Isherwood, Loring, and Emery, all proving, under certain conditions, the overwhelming influence of initial condensation in the steam-cylinder; but the writer is not aware that any one of them has established a rational and general law applicable in all cases where the piston and valves are proven tight.

In a lecture delivered at the Institute of Civil Engineers, May, 1883, Sir William Thomson used the following words: "In physical science, a first essential step in the direction of learning any subject is to find principles of numerical reckoning, and methods for practically measuring some quality connected with it. I often say, when you can measure what you are speaking about, and express it in numbers, you know something about it; but when you cannot measure it, when you cannot express it in numbers, your knowledge is of a meagre and unsatisfactory kind; it may be the beginning of knowledge; but you have scarcely in your thoughts advanced to the stage of science, whatever the matter may be."

It was with the feeling so well expressed by Professor Thomson that the writer undertook to find out, as well as he could, whether the Newtonian law of cooling was applicable to the action of steam inside the steam-cylinder.

So far as he knows, no writer, from the time of Carnot to this present date, has essayed to include all the data which must affect the condensation and expansion of steam in one general formula applicable to all cases.

The factors which must enter into such a formula will be seen to be too numerous to admit of the graphical treatment which, following the high authority of Rankine, quite a number of his followers have essayed, with varying degrees of approximation to correctness of result, and apparently quite oblivious of the fact that they are reasoning in a circle from empirical statement of experimental results, and that they are shut out completely from the hope of logically searching for the best attainable conditions in the use of steam.

Should it prove that this discussion of initial condensation is a scaffolding for the walls and roof of a structure which will house all the warring experimental proofs and mathematical discussions as to economy of steam, and that the hitherto unknown law of condensation of steam inside of the cylinder of a steam-engine has been formulated with practical accuracy, the writer will regard himself as most fortunate in having been able to complete the actual theory of the steam-engine at work.

A step in the right direction was made in locating the place and time of the condensation in steam-cylinders by M. Léloutre.

G. A. Hirn (*Théorie Mécanique de la Chaleur*, 1876, Vol. ii. page 55) says:

"In comparing the actual cost of our engines per stroke with the theoretic cost obtained by multiplying the volume opened to the steam by the density of this vapor supposed superheated, I always found the first result very much higher than the second. For a long time I continued to believe that this increase of cost arose from piston-leaks. These leaks seemed to me both natural and probable, since we were working (as I then believed) at a very high temperature, capable of destroying or evaporating the lubricants used in the cylinder to diminish the friction and wear of the parts.

"I yield to M. Léloutre the whole merit of having opposed these views, of having suggested to me as both possible and probable the existence of condensation of steam during admission resembling a loss by leakage, and giving to the steam a density much greater than would be obtained by calculation. I yield to him the merit of having in a manner forced me to make new experiments which would, by various methods, bring the truth to light."

In a controversy with Zeuner, Hirn denies the possibility

of establishing a general theory of the steam-engine, in the
following words :

" To-day my conviction remains as it existed twenty years
ago. A proper theory of the steam-engine is impossible.
An experimental theory based upon the motor itself in all
the forms in which it has been attempted in applied mechanics is alone able to lead to exact results."

The method of treatment used by the writer is only one
of many which have been relinquished because of giving
impossible results when laboriously elaborated from actual
data more or less deficient on important points.

This should not be construed as a complaint. The keen
personal delight arising from the discovery of new truth
would have compensated for far more toil had it been required.

So far as he knows, the writer's method is new, and certainly original, and has been adopted mainly for the purpose
of rendering the results intelligible to those who are actually
engaged in the designing and care of engines.

It has shown that the wide differences in experimental
results of tests of different types and sizes of engines are not
irreconcilable, and that the builder of small engines of the
non-condensing type is quite as right in adopting four expansions as the builder of enormous marine engines of the
compound type is in adopting expansions of ten or more
volumes.

The value of the constant of condensation is not so accu-
rately determined as we could wish, and will require further
experiment. It probably is very nearly right, and cannot
lead users astray in practice.

There is no test of an incomplete theory or superficial
reasoning like an experiment.

The mechanical skill of Messrs. Wheelock, Reynolds, and
Harris have rendered experiment possible, and the thorough

and careful work of Mr. Hill in making the experiments upon them has placed within our hands results which should accord with a proper theory.

The supplementary experiments of Messrs. Gately and Kletzsch, although not so accurate, add to the completeness and breadth of the proof of the law of condensation. They have "builded better than they knew," and deserve acknowledgment for their services and industry.

In making the calculations the steam-tables of Mr. Chas. T. Porter have been used. (The Richards Indicator.)

The author regrets the length of some of the formulæ, caused by the necessity of including all the various influencing conditions in one expression. However, they will be found to be very easily comprehended.

Many writers and experimenters have stated and proved the advantages of expansion, compression, superheating, steam-jacketing, high speeds over slow speeds, large engines over small engines, high pressures over low pressures, condensing over non-condensing engines, and of compounding cylinders; but through lack of a sufficiently thorough quantitative study of these expedients they have in many instances become partisan in their views, and have drawn incorrect general conclusions from their experiments and reflections.

It has been the wish of the writer by a careful quantitative weighing of the results to define the limitations of these various expedients and to enable others to see where and how they should be used.

In this departure from older paths the writer cannot feel quite sure of his ground on all points, although convinced of the practical accuracy of his theory for technical uses, nor does he lay claim to infallibility; he presents his facts, computations, and ideas for your consideration and verification, asking your assistance in a research of inestimable value to the arts, and assuming, not the attitude of one having authority, but only that of a patient student of the real facts of the action of steam inside of the steam-cylinder, as derived from and verified by the most reliable experiments upon the steam-engine accessible to him.

It has been thought best, though publishing this in a separate book-form, to page it for insertion in the third edition of THE RELATIVE PROPORTIONS OF THE STEAM-ENGINE, of which it properly forms one chapter.

W. D. M.

UNIVERSITY OF PENNSYLVANIA,
Philadelphia, 1886.

CONTENTS.

CHAPTER XVI.

THE LIMITATIONS OF THE EXPANSION OF STEAM.

ART.   PAGE

73. The Influence of Condensation Neglected 176
74. The Influence of Initial Condensation............................. 185
75. The Law of Condensation of Steam in the Steam-Engine..... 187
76. The Influence of Compression....................................... 192
77. The Constant of Condensation 195
78. The Influence of Superheating..................................... 196
79. The Influence of the Steam-Jacket 204
80. The Influence of the Valve Movement.......................... 204
81. Computation of Condensation....................................... 205
82. Probable Errors of the Table.. 208
83. The Influence of the Condenser..................................... 210
84. The Jet-Condenser... 211
85. The Surface-Condenser 211
86. The Law of Re-Evaporation after Cut-off 213
87. The Influence of Compounding Cylinders 215
88. The Relation of Cylinder Ratio to Ultimate Expansion...... 222
89. The Influence of the Receiver and Clearances 223
90. The Horse-Power of Compounded Cylinders...................... 227
91. The Influence of Cranks at Right Angles........................ 228
92. Condensation in Compounded Cylinders 234
93. Methods of Experimentation 236
94. Code for Test of Boilers.. 236
95. Code for Test of Steam-Engines.................................... 241

CHAPTER XVI.

THE LIMITATIONS OF THE EXPANSION OF STEAM.

(73.) The Influence of Condensation Neglected.

FIG. 32.

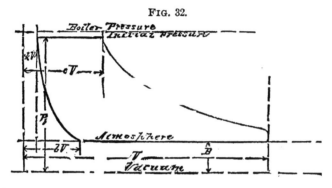

Let $(\mathbb{P})$ = the desired horse-power of the engine.

" P = the mean effective pressure of steam in cylinder in pounds per square inch.

" P_b = the absolute initial pressure of steam in pounds per square inch.

" B = the absolute back pressure of steam in pounds per square inch, exhaust open.

" e = the fraction of the volume of the steam-cylinder at which steam is cut off.

" b = the fraction of the volume of the steam-cylinder at which exhaust closes, being measured from the same end of the diagram from which e is measured.

" k = the fraction of the volume V of the steam-cylinder allowed for clearance.

" V = the volume of the steam-cylinder, inclusive of one clearance.

The best method of obtaining the volume of the steam-cylinder (V) and the clearance (k) is to fill the cylinder with water with the crank 'on the most distant dead point, and then to fill the clearance with water. The weight of water in the clearance, divided by the weight of water in the whole cylinder and clearance, will give the value of k. Reference to the diagram (Fig. 32) will make this clear.

The mean effective pressure =

$$P = eP_b \left[1 + \text{nat. log} \frac{1}{e} - \frac{k}{e} \right] - B \left[1 - b \left(1 - \text{nat. log} \frac{b}{k} \right) \right]. \quad (218)$$

For a cut-off e the water used per horse-power per hour is approximately, when we neglect the saving by compression,

$$W = \frac{859375\, e}{S \left\{ eP_b \left(1 + \text{nat. log} \frac{1}{e} - \frac{k}{e} \right) - B \left[1 - b \left(1 - \text{nat. log} \frac{b}{k} \right) \right] \right\}} \quad (219)$$

in which W = weight of water evaporated, and S = specific volume of the steam for the pressure P_b.

Formula (218) is stated under the assumption that the curve of expansion is an equilateral hyperbola, which we will show to be approximately an average of good indicator-cards.

The usual thing among many of the writers of the present day is to dismiss the isothermal curve with a condemnation and to make use of the adiabatic curve, but unfortunately for them it does not agree so well with the actual curve of pressures from our best engines as taken with an indicator as does the isothermal curve.

It will be observed that the curve of expansion used is the equilateral hyperbola. *This curve coincides, with great accuracy, with the best results derived from indicator-diagrams, and we can therefore safely use it until an adiabatic engine is produced for our discussion.*

Reference to the subjoined diagrams (Figs. 33, 34, 35, 36, 37, 38), which are all taken from engines of excellent workmanship not embarrassed by sluggish valve-motions, will convince the reader of the accuracy of this statement.

FIG. 33.

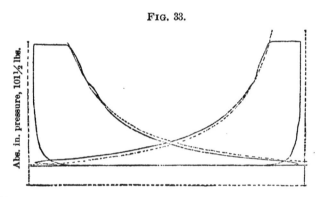

NON-CONDENSING HARRIS-CORLISS, NO. 14, JUNE 22, 7.45 A.M.   (J. W. Hill.)

FIG. 34.

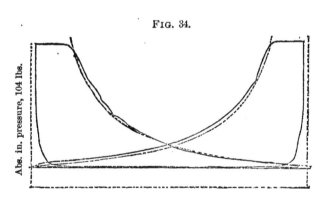

NON-CONDENSING HARRIS-CORLISS, NO. 28, JUNE 22, 11.15 A.M.   THOMPSON INDICATOR.

The broken curve is an equilateral hyperbola from the point of cut-off in each case of Harris-Corliss Engine. (J. W. Hill.)

Fig. 35.

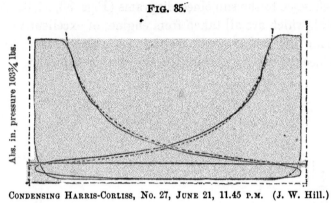

CONDENSING HARRIS-CORLISS, NO. 27, JUNE 21, 11.45 P.M. (J. W. Hill.)

Fig. 36.

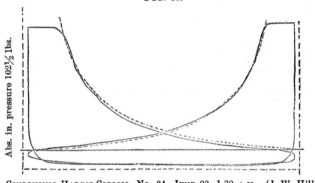

CONDENSING HARRIS-CORLISS, No. 34, JUNE 22, 1.30 A.M. (J. W. Hill.)

Fig. 37.

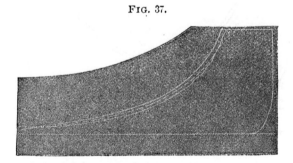

DIAGRAM OF THE BROWN ENGINE. NON-CONDENSING. (Taken by F. W. Bacon.)
The broken curve is the equilateral hyperbola. Absolute initial pressure, 90 lbs.

FIG. 38.

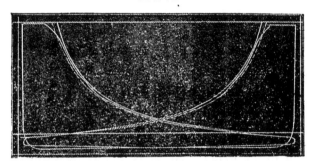

DIAGRAM OF BUCKEYE AUTOMATIC ENGINE. KORTING JET-CONDENSER.
The solid curve above is the adiabatic curve, and the broken curve below is the
isothermal expansion curve, traced on the diagram for comparison, and
beginning at the toe of the diagram. (J. W. Thompson.)
Absolute initial pressure, 93 lbs.

Equation (219), if we take account of the steam saved by
compression, becomes, when measurements are taken from
the diagram,

$$W = \frac{859375\left(e - k\dfrac{B_m}{P_c}\right)}{PS},\qquad (219\,a)$$

in which B_m = maximum pressure of compression,
   and P_c = the pressure at point of cut-off;
or, in a more general form, not quite so accurate as (219 a),

$$W = \frac{859375\left(e - b\dfrac{B}{P_b}\right)}{PS}.\qquad (219\,b)$$

This formula is more accurate than (219).
The Warrington-Thompson rule is,

$$W = \frac{859375\,(1 - b_1)}{PS_1},\qquad (219\,c)$$

in which b_1 is the fraction of the theoretical stroke (or volume inclusive of one clearance, V,) at which a line drawn

parallel to the atmospheric line, at a height equal to the terminal pressure of expansion, cuts the curve of compression. S_1 is the specific volume of the steam for its terminal pressure.

This very practical and simple rule had its origin as stated in the following letter from Mr. J. W. Thompson, Salem, Ohio:

"There was formerly in our employ a young man, by name Jesse Warrington, who was something of a prodigy in the way of quick, instinctive, and short-cut mathematics. Away back in the '70s, about 1874, if we mistake not, he suggested to the writer that there might be a constant figured out which, when divided by the product of the mean effective pressure and the volume of the total terminal pressure, would give the theoretical rate of water consumption, independently of any knowledge of the size and speed of the engine. Acting on the hint, I figured one out, it being obviously the consumption of water per indicated horse-power per hour of an engine subjected to one pound maximum efficient pressure, and driven by solid water instead of steam. The process was as follows: A horse-power being 33,000 foot-pounds per minute, it will be 23,760,000 inch-pounds per hour; this divided by 27.648, the number of cubic inches in a pound of water when a cubic foot weighs 62.5 pounds, gives 859,375, and I remember being somewhat struck with the singular fact that the calculation comes out exactly even, without a fraction. $\dfrac{33,000 \times 60 \times 62.5}{144}$ also gives it."

If the point of cut-off is easy to determine, as in the diagrams of properly-constructed engines already given, formulæ (219), (219 *a*), or (219 *b*) may be used as the judgment may dictate.

Should it be impossible to determine the point of cut-

16

off, then the Warrington-Thompson rule (219 *c*) must be used.

*P*lacing an indicator on the steam-chest will show, by a sudden jog in the line, the point of cut-off, or reveal defects or sluggishness of valve-motions. See Fig. 39.

The quantity of steam recorded by the diagram, if deduced by the Warrington-Thompson rule, is simply the amount of steam remaining as vapor in the steam-cylinder after the steam has done its work, and with all its conveyed heat it is immediately thrown away through the exhaust-port.

The quantity $\dfrac{Bb}{P_i}$, equation (219 *b*), is quite small in a non-condensing engine, *and should never exceed the clearance k in value.* In a condensing engine it is usually very much smaller. In making close calculations of the water used, we cannot afford to neglect this quantity; but it does not sensibly affect the point of cut-off.

It is not possible to avoid initial loss of heat, to construct an engine without clearance, or to obtain a perfect vacuum; but if, for the purpose of obtaining an extreme limit, we neglect these, equation (219) becomes

$$W = \frac{859375}{SP_b\left(1 + \text{nat. log } \dfrac{1}{e}\right)} \text{ pounds per H.-P. per hour.}$$

The following table, therefore, gives a minimum and unattainable quantity of dry saturated steam for each atmosphere up to 12, and each expansion up to 12. The fraction of its volume at which the steam is cut off will be found on the upper horizontal line of the table, and the steam-pressures on the vertical line.

Table of Weight of Dry Saturated Steam required per Horse-Power per Hour, under Assumptions giving the Minimum Value (Initial Condensation being neglected).

Pressures in Atmospheres.	Pressures in Lbs. per Square Inch (Abs.). P_0	$\dfrac{1}{1+\text{nat. log }\frac{1}{e}}\cdot\dfrac{1}{P_0 S}$ True Point of Cut-off.	1	$\frac{1}{2}$	$\frac{1}{3}$	$\frac{1}{4}$	$\frac{1}{5}$	$\frac{1}{6}$	$\frac{1}{7}$	$\frac{1}{8}$	$\frac{1}{9}$	$\frac{1}{10}$	$\frac{1}{11}$	$\frac{1}{12}$	S Specific Volumes.
			1.000	0.5907	0.4764	0.4191	0.3833	0.3582	0.3394	0.3248	0.3128	0.3028	0.2943	0.2870	
1	14.7	.0004135	35.54	20.99	16.93	14.90	13.62	12.73	12.06	11.54	11.11	10.76	10.46	10.20	1645.
2	29.4	.0004034	34.67	20.48	16.51	14.53	13.29	12.42	11.77	11.26	10.84	10.50	10.20	9.95	843.1
3	44.1	.0003951	33.94	20.06	16.18	14.23	13.02	12.16	11.53	11.03	10.62	10.28	9.99	9.74	572.
4	58.8	.0003894	33.47	19.77	15.95	14.03	12.83	11.99	11.36	10.87	10.47	10.13	9.85	9.60	436.7
5	73.5	.0003833	32.94	19.46	15.69	13.81	12.63	11.80	11.18	10.70	10.30	9.97	9.69	9.45	354.9
6	88.2	.0003777	32.46	19.17	15.47	13.60	12.44	11.63	11.02	10.54	10.15	9.83	9.55	9.32	300.2
7	102.9	.0003728	32.03	18.92	15.26	13.43	12.28	11.48	10.87	10.41	10.02	9.70	9.43	9.20	260.7
8	117.6	.0003670	31.54	18.63	15.03	13.22	12.09	11.30	10.71	10.25	9.87	9.55	9.28	9.05	231.7
9	132.3	.0003615	31.06	18.35	14.80	13.02	11.91	11.13	10.55	10.09	9.72	9.41	9.14	8.92	209.1
10	147.0	.0003565	30.64	18.10	14.60	12.84	11.75	10.97	10.40	9.95	9.58	9.28	9.02	8.79	190.8
11	161.7	.0003519	30.23	17.86	14.41	12.68	11.59	10.83	10.27	9.82	9.46	9.16	8.90	8.68	175.8
12	176.4	.0003472	29.83	17.63	14.22	12.50	11.44	10.69	10.12	9.69	9.33	9.04	8.78	8.56	163.3

Inspection of this table will show how rapid the increase of economy due to expansion appears to be up to about 10 expansions, where gain by this expedient becomes comparatively small with subsequent increase.

The first vertical column shows also the slower gain due to increased pressures up to 12 atmospheres; this gain is of great practical value, as the steam appears to be drier at high pressures and the engine becomes more powerful for a given volume of cylinder.

The formula (217) given in the preceding chapter will give the uttermost limit attainable under the assumptions made for this table.

It is easy to see that so long as steam expands according to Marriotte's law, when acting inside of a steam-cylinder, and we are restricted to moderate pressures, we cannot hope to obtain a horse-power per hour for less than one pound of coal.

The cost of power may be divided into two accounts:

The first is the constant charges, which are—

(1) Interest on, deterioration of, and repairs to engine.

(2) Wages of firemen and engineer.

(3) Cost of oil and waste.

(4) Interest on, deterioration of, and repairs to boilers.

(5)   "   "   "   shelter if separate.

(6) Taxes and insurance on engine and boilers.

The second is the cost of the fuel and water required to produce and also condense the steam per horse-power per hour.

The constant charges may within narrow limitations be regarded as proportional to the power required. As the power required becomes larger the proportion between the constant charges and the power becomes smaller, provided we obtain it from one engine of not too expensive construction or size.

(74.) The Influence of Initial Condensation.—The cost of fuel and water required is divided into two parts:

(1) The amount of steam appearing from the diagram.

(2) The amount of steam lost by condensation in the steam-cylinder.

The losses in heating the water in the boiler and in conveying the steam to the engine are not properly chargeable to the engine, and, being an item for which separate provision must be made, will not be regarded.

The intending purchaser must first fix upon that size, speed, and type of engine which will give him the power he requires for the least cost in steam, and then obtain that engine for the least sum possible.

Cases may and do occur in which the different types of engines vary enormously in cost.

In such cases there is but one way to decide. A detailed calculation of the constant charges and the cost of steam must be made for each engine, and that engine showing the least sum per horse-power per hour will be the proper one.

It may occur that greater cost of steam per horse-power per hour will be recouped by a lesser constant charge per horse-power per hour.

It is never safe to guess, and particular attention should be paid to the cost of condensation of steam when a condenser is used.

We can then write the following equation for discussion:

$$Y = \frac{\text{Work}}{\text{Cost of work in steam}},$$

or, for one stroke of the engine,

$$Y = \frac{\text{Mean effective pressure} \times \text{Volume of steam-cylinder}}{\text{Cost in [Indicated steam + Initial condensation + Final condensation of steam]}}.$$

The final condensation of the steam ordinarily requires

16 *

some of the power in the engine, and demands an additional outlay for condenser and a large quantity of water.

Where water is costly, this may result in a decision in favor of a non-condensing engine.

The whole matter of final condensation is one for a calculation and comparison subsequent to the preliminary calculations of the steam required per horse-power per hour as delivered by the boiler.

We can then reduce the above equation to the form,

$$Y = \frac{\text{Mean effective pressure}}{\text{Indicated steam} + \text{Initial condensation}}.$$

The indicated steam shown by the diagram to exist as vapor in the engine-cylinder has already been discussed at length.

There remains for our consideration the initial condensation.

In order to establish a clear understanding, let us carefully follow the steam in the cylinder through one stroke.

The order of events would seem to be as follows, the engine having attained its regular speed, and the cylinder an average heat:

The entering steam touching the interior of the cylinder condenses very rapidly and warms it up to the temperature of the steam; this warmth proceeds to a depth proportional to the depth already cooled by the exhaust; the steam then expands after cut-off, falling in temperature and losing heat, first by warming up the cooled cylinder-walls, secondly in doing work; however, the heated iron of that part of the cylinder exposed before cut-off gives up heat and vaporizes the condensed water of initial condensation in the attempt to equalize the temperatures throughout the cylinder, which is effected by a transfer of condensation following the motion of the piston-head.

At the end of the stroke the temperature of the whole internal *surface* and of the steam is that of the terminal pressure, the steam having really expanded with fresh accessions of heat and of vapor from that part of the cylinder exposed to initial steam; that is, exposed before the cut-off occurs.

Next in the order of events the exhaust opens and the whole interior of the cylinder is exposed to the temperature of the exhaust, the piston and cylinder-head being exposed on an average twice as long as the cylinder-walls.

In every engine these changes, whatever they may be, establish an equilibrium among themselves, and the result is that a certain uniform quantity of heat is lost at each stroke, provided the thermal value of the steam does not vary.

As long as we know the equilateral hyperbola to be an average of the curves produced by good indicators on good engines, the condensation of steam during expansion or its re-evaporation is of minor importance, and we can return to the initial condensation of steam for the present.

It is nonsense to discuss the curve of adiabatic expansion until we can produce adiabatic engines.

The first step in the quantitative investigation of the condensation of steam is the establishment of a standard or constant.

We will assume this to be THE WEIGHT OF STEAM CONDENSED PER MINUTE IN RAISING THE TEMPERATURE OF A SURFACE OF CAST IRON OF ONE SQUARE FOOT AREA ONE DEGREE FAHRENHEIT.

This weight can be converted into heat-units by multiplying it by the total heat of steam less heat of its water for the given pressure at cut-off.

(75.) **The Law of Condensation of Steam in the Steam-Engine.**—The loss of heat by the steam in the cylinder is proportional to—

I. The difference of temperatures of the steam at the point of cut-off and while being exhausted.

II. The area of cast iron exposed to the entering steam up to the point of cut-off.

III. The time of exposure of the interior surface of the steam-cylinder to the exhaust steam.

IV. It is reduced by compression subject to the same laws, but, as this is quite a small quantity in most cases, we will neglect it for the present.

To the notation already in use let us add :

T_b = temperature of steam at point of cut-off.

T_e = temperature of steam during exhaust.

N = number of strokes of engine per minute.

C = the constant of condensation.

The first factor $(T_b - T_e)$ can at once be written.

Assuming the crank to have uniform rotation, and the angle which it forms with the centre line of the steam-cylinder, measured from the end from which it is retreating, to be a.

If we take account of the variable time of exposure of the cylindrical walls of the cylinder, as also of the variable value of the area of the cylindrical elements, we have for the time of exposure to exhaust, multiplied by the area,

For the piston-head and the cylinder-head, and the clearance which is neglected,

$$\frac{\pi d^2}{2N}.$$

For the variable area for each increment of the angle a, the crank being assumed to have a uniform rotary motion,

$$(\pi d)ds = + (\pi d)\frac{s}{2} \sin a \, da.$$

For the variable time $\dfrac{a}{180N}$ of exposure to exhaust of each element, $(\pi d)ds$. For their product

$$\frac{(\pi d)s}{2N} \frac{a}{180} \sin a \, da.$$

For the product of time and area of interior of cylinder, between limits,

$$\frac{\pi d}{2N}\left[d+\frac{s}{180}\int_{\substack{a=180° \\ a=\cos^{-1}(2e-1)}}^{a=180°} a\sin a\,da\right],$$

but, $\int a\sin a\,da = \sin a - a\cos a = \sqrt{1-\cos^2 a} - a\cos a,$

for $a = 180°$ we have $\pi,$

for $a = \cos^{-1}(2e-1)$ we have

$$\sqrt{1-(2e-1)^2}\,(2e-1)\cos^{-1}(2e-1).$$

Therefore we have, as a final quantity for any cut-off $e,$

$$\frac{\pi d}{2N}\left[d+\frac{s}{\pi}\{\pi-\sqrt{1-(2e-1)^2}+(2e-1)\cos^{-1}(2e-1)\}\right].$$

This latter expression should be multiplied by $(T_b - T_e)$ and the constant of condensation $C,$ giving for the total condensation during one stroke and without compression,

$$C\left[\left(T_b-T_e\right)\frac{\pi d}{2N}\left\{d+\frac{s}{\pi}\left\{\pi-\sqrt{1-(2e-1)^2}+(2e-1)\cos^{-1}(2e-1)\right\}\right\}\right]. \quad (220)$$

We also have for the combined time and area of exposure of initial steam, or putting b for e of compressed steam,

$$\frac{\pi d}{2N}\left\{\frac{d}{\pi}\cos^{-1}(1-2e)+\frac{s}{\pi}\left[\sqrt{1-(1-2e)^2}-(1-e)\cos^{-1}(1-2e)\right]\right\};$$

but this is a lesser and not the controlling quantity.

Tracing the curve of condensation,

$$Y=1-\frac{1}{\pi}\left\{\sqrt{1-(2e-1)^2}+(2e-1)\cos^{-1}(2e-1)\right.$$

If in this equation we make e the independent variable, it represents a long, easy curve which approximates very closely to a straight line from $e = 0.15$ to $0.5.$

We have, as shown in the diagram, values of Y as ordinates for each abscissa e to unity.

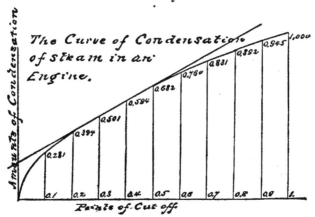

Between the limits $e = .15$ to $e = .55$ the equation of the nearly coincident straight line is

$$Y = e + 0.194.$$

Equation (220) becomes

$$C\left[\left(T_b - T_e\right)\frac{\pi d}{2N}\left(d + Ys\right)\right], \qquad (221)$$

or between the above limits for e approximately,

$$C\left[\left(T_b - T_e\right)\frac{\pi d}{2N}\left(d + \{e + .194\}\, s\right)\right]. \qquad (221\,a)$$

The weight of steam in the form of vapor at the moment of cut-off is $\dfrac{62.5}{S}\left(e - k\dfrac{B_m}{P_b}\right)V = W.$ \qquad (222)

Neglecting losses due to pipe-condensation and priming, and the gain due to compression, and adding to this the weight of condensed steam in the cylinder at the moment of cut-off, we have $\dfrac{62.5}{S}\,eV + C\left[\left(T_b - T_e\right)\dfrac{\pi d}{2N}\left(d + Ys\right)\right] = W_a,$

or, between limits set above, $Y = e + .194.$

$$\frac{W_a}{W} = r = 1 + \frac{SC^-}{62.5N}\left(T_b - T_e\right)\frac{2}{e}\left(\frac{1}{s} + \frac{e+.194}{d}\right), \quad (223)$$

in which d and s are measured in feet.

$$\text{Let } A = \frac{62\frac{1}{2}}{S} \text{ and } D = 2\,\frac{T_b - T_e}{N}\,C,$$

We have $r = 1 + \dfrac{D}{A}\dfrac{1}{e}\left(\dfrac{1}{s} + \dfrac{e+.194}{d}\right).$ $\qquad$ (223 a)

These equations give us a means of determining from the indicator-card the steam actually furnished by the boiler.

Presuming the value of C to be known, and dry saturated steam used, we can predict the most economical expansion for any size of cylinder, number of strokes, pressure of initial steam, and back pressure during exhaust.

If we desire to know the practically most economical number of expansions, or its inverse, the true point of cut-off, we must find the maximum of the following equation:

$$y = \frac{PV}{r\dfrac{62.5}{S}eV} = \frac{P}{Are}.$$

Neglect clearance and compression in the value of P, and substitute for r its value, we have

$$y = \frac{eP_b\left(1 + \text{nat. } \log\dfrac{1}{e}\right) - B}{Ae + D\left(\dfrac{1}{s} + \dfrac{e+.194}{d}\right)}.$$

Differentiating with respect to e for a maximum, we have

$$e = \frac{B}{P_b} + \left(\frac{1}{s} + \frac{.194}{d}\right)\frac{Dd}{Ad+D}\,\text{nat. } \log\frac{1}{e}. \quad (224)$$

This most important transcendental equation must be solved tentatively; or we can assume a value for e which must always be greater than $\dfrac{B}{P_b}$, and deduce the most economical ratio of stroke to diameter of steam-cylinder. This equation is not exact when e is less than .15 or exceeds .50, which is not a usual thing.

(**76.**) **The Influence of Compression.**—If we take account of clearance and compression, we have

$$y = \frac{eP_b\left(1+\text{nat. }\log\dfrac{1}{e}-\dfrac{k}{e}\right) - B\left(1 - B\left[1 - \text{nat. }\log\dfrac{b}{k}\right]\right)}{A\left(e - b\dfrac{B}{P_b}\right)\left(1+\dfrac{D}{A}\dfrac{1}{e}\left\{\dfrac{1}{s}+\dfrac{e+.194}{d}\right\}\right)}.$$

Differentiating for a maximum, we have

$$e = \frac{B\left[1 - b\left(1 - \text{nat. }\log\dfrac{b}{k}\right)\right]}{P_b} + k + \frac{D\left(\dfrac{1}{s}+\dfrac{.194}{d}\right) - \dfrac{ABb}{P_b}}{A + \dfrac{D}{d} + \dfrac{BDb}{e^2 P_b}\left(\dfrac{1}{s}+\dfrac{.194}{d}\right)}\;\text{nat. }\log\dfrac{1}{e}.\quad(224a)$$

Assuming b and $k = 0$, we obtain equation (224).

Use equation (224) to obtain an approximate value of e, and substitute in (224 a) to obtain an accurate value when clearance and compression are to be considered.

The compression of the steam by early closing of the exhaust-port has two effects: it diminishes the power of the engine and increases its economy by saving, at each stroke, an amount of steam which fills the clearance at the maximum compression pressure B_m. May it not also diminish the amount of initial condensation of steam by covering the surfaces of the piston- and cylinder-heads with films of water at a temperature determined by the maximum compression pressure of saturated steam and the time of exposure? The barrel of the cylinder, near the end, may also have a preliminary warming, due to the hyperbolic increase of pressure of the steam from the moment of exhaust-closure.

The following pair of diagrams will illustrate (Fig. 39) the case just stated.

FIG. 39.

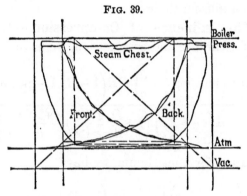

Porter-Allen Engine. Post-Office Building, Philadelphia, March 30, 1884. Scale,
40 Pounds per Inch.

Steam not tested, but not superheated.

Valves and piston not tested.

Engine non-condensing.

Stroke, 24″ = 2 ft.

Clearance, 4½ per cent. of stroke = 0.09 ft. deduced from expansion curve.

Diameter, 14½″ = 1.208 ft.

Abs. initial pressure, back end, 87 lbs.

  " " " front " 89 "

  ' " " mean, 88 "

  " back pressure at midstroke, 16 "

Temperature of initial steam, 318.45° Fahr.

Specific vol. of " " 300.8.

Temperature exhaust " 216.29° Fahr.

Number of strokes, 400 per minute.

Mean eff. pressure, front end = 20.62 ⎫

 " " " back " = 17.24 ⎬ Amsler's *P*olar. Pla-

 " " " mean = 18.93 ⎭ nimeter used.

Indicated horse-power = 74.91.

17

Maximum back pressure = pressure at cut-off.

(e) Point of cut-off, front end } = 0.168 mean value.
"   "   " back   " }

(k) Clearance (true),   0.043.

Cut-off less compression $(e-k)$,   0.125 = ⅛.

Indicated steam, $\dfrac{859375}{300.8 \times 18.93 \times 8} = 18.86$ lbs. per horse-power per hour.

Let T_c = the temperature Fahrenheit due to maximum compression pressure, then formula (220) becomes, when we neglect the time of exposure to compressed steam,

$$C\left[\left(T_b - T_c\right)\frac{\pi d^2}{2N} + \left(T_b - T_e\right)\frac{ds}{2N}\left\{\pi - \sqrt{1-(2e-1)^2} + (2e-1)\cos^{-1}(2e-1)\right\}\right], \quad (220b)$$

or, simplifying as before,

$$C\left\{\left[T_b - T_c\right]d + \left[T_b - T_e\right](.194+e)s\right\}\frac{\pi d}{2N}. \quad (220\ c)$$

If the maximum compression pressure becomes equal to the pressure at cut-off $T_b = T_c$, assuming pressures and temperatures to vary together, giving for the condensation within .50 per cent. cut-off

$$C\left[T_b - T_e\right]\frac{(.194+e)\pi ds}{2N}. \quad (220\ d)$$

Equation (223), using the value given for condensation by (220 c), becomes

$$r = 1 + \frac{SC}{62\frac{1}{2}N}\left[\left(T_b - T_c\right)\frac{2}{es} + \frac{2(e+.194)}{e}\frac{}{d}(T_b - T_e)\right], \quad (223\ b)$$

or, if maximum compression pressure equals pressure at cut-off,

$$r = 1 + \frac{SC}{62.5N}\left(T_b - T_e\right)\frac{2(e+.194)}{ed}, \quad (223\ c)$$

or, $\qquad r = 1 + \dfrac{D}{A}\dfrac{(e + .194)}{de}.$ $\qquad (223\,d)$

$$e = \frac{B\left[1 - b\left(1 - \text{nat. log } \dfrac{b}{k}\right)\right]}{P_b} + k + \frac{.194\,\dfrac{D}{d} - Ak}{A + \dfrac{D}{d} + .194\,k\,\dfrac{D}{de^2}}\ \text{nat. log } \frac{1}{e}\ ; \quad (224\,b)$$

$k = \dfrac{Bb}{P_b}$, and the factor $\dfrac{1}{s}$, disappears from formula. $\quad$ (224 a)

If compression to initial pressure does prevent condensation by the piston and cylinder-heads, (223 d) is a minimum value.

It is probable that in most cases where clearance and compression are slight the saving of condensation is small from compression, owing to lack of time. In cases of large clearances and early compression the time of exposure is sufficient to enable warming of piston and cylinder-heads by condensation of the compressed steam.

(77.) **The Constant of Condensation.**—In order to prove the value of C to be constant under all conditions existing in steam-engines, we must refer to practical experiment.

In its most exact form we can, from equations (223) and (220 b), write

$$C = \frac{62\tfrac{1}{2}\left(e - k\dfrac{B_m}{P_b}\right)\left(r - 1\right)N}{2S\left[\left(T_b - T_c\right)\dfrac{1}{s} + \left(T_b - T_e\right)\dfrac{Y}{d}\right]}. \qquad (225)$$

For a cut-off not greater than .50 or less than .15, we have, approximately,

$$C = \frac{r - 1}{\dfrac{2S}{62\tfrac{1}{2}eN}\left[\left(T_b - T_c\right)\dfrac{1}{s} + \left(T_b - T_e\right)\dfrac{e + .194}{d}\right]}. \qquad (225\,a)$$

Without compression being used,

$$C = \frac{r-1}{\frac{2S}{62\frac{1}{2}eN}\left[\left(T_b - T_e\right)\left(\frac{1}{s} + \frac{e+.194}{d}\right)\right]}. \qquad (225\,b)$$

Unless the compression of the steam is very marked, it will have little effect in reheating the piston- and cylinder-heads, and the exhaust temperature T_e must be used. With even slightly leaky steam-valves, the compression line is mainly due to a drizzle of steam into the cylinder after the exhaust-port is closed, and to the lead, and not to the steam entrapped by the closing of the exhaust-ports.

With compression to initial pressure and large clearance we may have more nearly

$$C = \frac{r-1}{\frac{2S}{62\frac{1}{2}eN}\left[\left(T_b - T_e\right)\left(\frac{e+.194}{d}\right)\right]}. \qquad (225\,c)$$

(78.) The Influence of Superheating.—As some of the experiments which will be discussed in a subsequent table were said to have been made with superheated steam, it will be necessary to form an opinion as to its action inside of the steam-cylinder.

This is a difficult undertaking, and the reader must use his own judgment in deciding whether or not to adopt the opinions put before him.

Superheating steam has two consequences. The temperature of the steam is increased beyond its temperature of saturation, and its specific volume is increased at the same time.

The behavior of the iron of the cylinder, as well as the mechanical conditions of the introduction of the steam into it, will not permit us to accept the results of calculation under other circumstances than those actually existing in the engine itself.

The following suggestions will reveal why we cannot expect superheated steam to act in an iron cylinder as it would in an adiabatic vessel.

If the steam condenses only as it comes in contact with the walls,—that is, condenses as a piece of ice melts, on the outside,—the main body of the steam *will remain superheated,* and successive layers of steam on the outside be condensed, until the walls are of the same temperature as its temperature of saturation, after which the superheated steam would strive to re-evaporate the condensation, and the iron to keep it condensed and add more to it, with all the advantage on the side of the iron. Since, if we assume steam of an average specific volume of 300, their comparative weights, volume for volume, are,—iron, 2160; steam, 1; and as the specific heat of steam is 0.48, and that of iron 0.12, we see that, volume for volume, or layer for layer, iron will take 540 times as much heat as steam. So we see the iron will not only keep what it has condensed, but will add more to it by conducting the heat from the film of water upon it away so rapidly as to cause additional condensation in the adjacent layers of steam, which have comparatively no conducting power at all, if we can draw any inference from the ease with which priming occurs in steam, and the suspension of globules of water during expansion of steam.

The advantages of superheated steam will be found to be in the more complete re-evaporation during expansion leaving a drier cylinder, and consequently less to be evaporated by the heat of the cylinder during the exhaust, which will therefore be cooled to a lesser depth.

It is not surmised that the temperature of the iron of the steam-cylinder rises above the temperature of saturated steam of the initial pressure; but the possibility is sug-

17 *

gested that the iron gets *more* heat in proportion to the superheating, and therefore re-evaporates the steam more readily and thoroughly. Certainly it would not appear, in the table of the Harris-Corliss engine condensing, that 66 per cent. of the indicated steam should be condensed at cut-off, or with the same engine non-condensing 56 per cent. of the indicated steam should have been condensed at cut-off, if all the superheating of the initial steam had not vanished before expansion began.

Neither does it appear from the diagrams that the super-heating had the effect of raising the expansion curve above an equilateral hyperbola in the Harris and other engines, nor does wet steam seem greatly to modify the expansion curve of the Brown and Porter-Allen engines.

The fact, however, is undeniable that the amount of heat conducted away is proportional to the conductivity and heat-capacity of the iron cylinder, and we are told that this loss is directly proportional to the difference of temperatures.

It is worthy of note that the superheated steam in the Harris-Corliss and other engines must have instantly become very wet steam upon its contact with the cylinder; and therefore if damage is done by superheated steam, its evil effects will appear upon the valve-motion and upon lubricants fed through the steam-pipe.

It would seem as if especial care given to the jacketing of the cylinder-barrel would be almost useless trouble until we have suppressed initial condensation *due mainly to the piston- and cylinder-heads.*

Even were we to obtain very good non-conducting surfaces for these latter parts, it is doubtful if a film of water would not appear on them and demand conversion into steam at the beginning of each stroke.

Possibly the greater economy of actual steam which it is claimed uniformly results from superheated steam lies in the certainty of *dry* saturated initial steam, the greater certainty of more complete re-evaporation during expansion, and the presence of less watér on the interior walls of the steam-cylinder demanding re-evaporation during exhaust than would occur with saturated steam.

Practically, superheating is often done with the waste gases of combustion, and therefore costs nothing for fuel. Because what is called saturated steam is not always dry, we should not jump to the conclusion that under all circumstances it is the best plan to use superheated steam.

Let us essay to follow the interaction of superheated steam and the iron cylinder walls from the moment of its entrance to the point of cut-off.

At the moment of entering, the superheated steam fills the clearance, forming a thin disk of vapor between two circular iron plates (the cylinder- and piston-heads) of many degrees lower temperature. As it touches the iron it first parts with its superheat, and then condenses copiously upon these surfaces, and, so long as the pressure is not lessened, the iron will not permit re-evaporation, although the surfaces may be streaming with water.

Hirn (Vol. ii. page 62, *Théorie Mécanique de la Chaleur*) thus describes the action of superheated steam:

"In a cylinder, at the moment of admission, it is possible, and it ought, to stream with water upon the walls; whilst the free space should be filled with steam of 231 Centigrade degrees (the temperature of his superheated steam).

"The condensation during the admission, as well as the evaporation of a part of the water during expansion, takes place *by contact, part by part,* and not by the cooling or heating of *all* the mass present."

From the narrow, long orifice of the port at the side of the cylinder this steam passes in, frequently at the rate of a mile a minute, and must create strong swirls and eddies, thus giving the superheated steam opportunity to come in contact with the streaming walls and part with its superheat. The slow motion of the piston at the beginning of the stroke favors this action, for it requires one-third of the time of the stroke for the piston to reach one-fourth of its travel.

The tendency of the superheat is to re-evaporate the moisture of the walls; and of the iron, to take the heat from this film of water, and the iron probably does it before it can re-evaporate. Perfect quiescence, as of melting ice, is necessary, to accept the statement of Hirn as regards the entire retention of superheat by steam.

It is more than probable that superheated steam will part with its extra heat to the water wherever it comes in contact with it. The exact action of the film of water interposed between the steam and the iron cannot be stated, but it is probably a very good non-conductor of heat to the iron from steam when saturated.

The expressions (220) and (220 a) give the amount of heat demanded by the iron of the cylinder before it will permit the steam to do its work.   .

The advantage of superheated steam lies in this: First. Superheating costs little additional fuel. Second. The iron takes just as great a quantity of heat as with saturated steam of the same pressure; but the superheat satisfies this requisition in part, and less weight of saturated steam is required to drive the piston ahead.

Zeuner, in his " Theory of Superheated Steam," *Zeitschrift des Vereins Deutsches Ingenieure*, Band XI., 1866, gives the following formula for the specific volume of superheated or saturated steam for any absolute temperature, *T*, in Centigrade degrees, and any pressure, *p*, in atmospheres:

v = the specific volume, metric system.

$$pv = B\,T - C\,p^{\frac{1}{4}}.$$

B and C are constants:

$$B = 0.0049287 ; \qquad C = 0.187815.$$

This empirical formula seems to hold equally good for the specific volume of both saturated and superheated steam, giving very close results in either case.

Therefore, denoting by V_1 and T_1 the specific volume and absolute temperature of superheated steam of a pressure p, we have

$$\frac{V_1}{V} = \frac{T_1 - \frac{C}{B}\sqrt[4]{p}}{T - \frac{C}{B}\sqrt[4]{p}}. \qquad (226)$$

If we transform this formula for the ratio of the specific volumes into Fahrenheit degrees and pounds pressure (absolute) per square inch, we have

$$S_s = S \frac{T_s + 459.4 - 35.022 \sqrt[4]{P_b}}{T_b + 459.4 - 35.022 \sqrt[4]{P_b}}. \qquad (226\,a)$$

In this formula, S_s is the specific volume of superheated steam at the pressure P_b, and T_s is the temperature of the superheated steam in degrees Fahrenheit.

T_s can be obtained by direct measurement of the temperature in the steam-pipe, or can be deduced with sufficient accuracy by multiplying its excess of thermal units over saturated steam by 2, and adding this to the temperature of saturated steam.

We have mentioned the extreme avidity of the absorption and emission of the heat of the steam by the interior surface of the cylinder. We should also recollect that strong currents, swirls, and eddies exist in the steam as it enters. Hirn assures us that it is possible for the steam not in contact with the iron walls to remain superheated, and this may be true wholly or only in part. *We believe it to be saturated steam.*

Equation (220 *b*), if we take *C* in British units and multiply it by 2, under the assumption that the specific heat of steam is 0.5, will give the degrees of superheat required, so that the heat abstracted by the iron may be met, and the case be equivalent to the use of saturated steam in an adiabatic cylinder, without this preliminary loss of heat.

There is a preliminary condensation and a mass of more or less superheated steam inside at the point of cut-off. There must be a more rapid re-evaporation during expansion, and a less amount of water for re-evaporation during exhaust. It is currently believed that superheating tends to score the cylinder, by causing such thorough re-evaporation during exhaust as to leave the cylinder dry and hot. Hirn fixes the safe temperature of superheated steam at 446° Fahr. It is easy to see, however, from equation (220 *b*), that differing proportions, pressures, and speeds require different amounts of superheat.

Hirn, since 1850, has devoted his talents as an experimenter and his profound knowledge of practical thermodynamics to experimental verification of thermodynamic laws.

The following *résumé* of his results is taken from the work of M. Marcel Deprez (*Rev. Univ. des Mines*, 1874):

(1) " A compound Woolf engine, worked for fifteen years with saturated steam, consumed 12¼ kilogrammes of steam per horse-power per hour (French). Suppressing the first cylinder and working with superheated steam expanded four times, the steam-consumption fell to 10 kilogrammes.

(2) " In another engine without jacket, with a single cylinder and 4 expansions, the consumption was 14.75 kilogrammes with saturated steam, and fell to 10 kilogrammes with the steam superheated to 250° Centigrade.

(3) "The two cylinders of the Woolf engine having been replaced by a single cylinder without jacket, the pressure being 4½ atmospheres, and with 4 expansions, the consumption fell to 8.53 kilogrammes.

(4) "In a high-speed engine (93 revolutions per minute), pressure 3¾ atmospheres, 4 expansions, steam superheated to 250° Centigrade, the consumption fell to 8.2 kilogrammes per horse-power per hour.

(5) "In an engine, pressure 5 atmospheres, 6 expansions, superheated to 245° Centigrade, the consumption was 7½ kilogrammes per horse-power per hour.

"In Number 4, the throttle being almost closed and the expansion very little, the consumption was 9.2 kilogrammes. Number 5, under the same conditions, consumed 10 kilogrammes.

"All these results were obtained for long periods; the power was measured by a brake; all were condensing engines, the last three having four valves; all were without steam-jackets, because M. Hirn believed superheating rendered them useless."

We see that M. Hirn attained by these expedients a greater economy, with very moderate pressures and expansions, than is usual in practice, by preventing transmission of heat through the cylinder-walls and by preventing initial condensation.

Hirn declares that improvement rendering the use of superheated steam safe and practicable is the greatest possible advance in steam-engine economy.

We cannot agree with Hirn that steam-jackets are useless in all cases,—they are at times and in some places necessary to the highest economy,—or that proper compounding is not beneficial.

The proof in two cases that throttling does not seriously affect the consumption—which was probably increased by the absence of expansion only, and prevented from becoming very great by the reduced initial pressure and temperature of the entering steam—is of very great value as affecting the relative economy of automatic cut-off and plain slide-valve engines with a throttling governor. Indeed, there is very little doubt that, in the case of small cylinders, quite as great economy, if not a greater, is as often attained with the throttling governor as with an automatic cut-off.

(79.) **The Influence of the Steam-Jacket.**—The steam-jacket should surround the whole of the cylinder at sides and ends. Except by means easily deranged, it seems hardly possible to introduce steam into the piston-head.

There is a steady progress of heat from the jacket to the interior surface of the cylinder, having the effect of diminishing the initial condensation, facilitating the re-evaporation during expansion, and increasing the amount of heat lost through the exhaust-port.

Diminishing the amount of water condensed is the surest means of preventing loss by re-evaporation during exhaust.

Compared with very bad engines, the extraordinary precautions taken in adding and covering a jacket quite frequently produce a great saving of steam; but it is very doubtful if any very great gain is made in the case of well-covered engines.   •

Well-authenticated cases of a saving of 3 to 5 per cent. are cited; but that error is more than probable in the use of indicators during a test, and may have occurred.

The sum total of this action of the jacket would seem favorable to economy, as, with a dry cylinder, the steam in the jacket can lose only by radiation, a comparatively slow process.

(80.) **The Influence of the Valve-Movement.**— Where the plain D-valve with three ports is used, condensation occurs in the steam-chest, the exhaust steam passing through the hollow, taking heat from it, which, in turn, is supplied from the "live" steam in the steam-chest.

Single slide-valves for two-cylinder engines prove frequently to be the cause of large losses which are unsuspected.

It would seem to be a pretty well established fact that, wherever steam economy is the first consideration, four-ported cylinders should be used, and every possible means to prevent iron surfaces from being alternately exposed to higher and lower temperatures of steam.

The advantage resulting from compression, superheating, steam-jacketing, and compounding steam-cylinders will be much more apparent in small than in large cylinders; because in small cylinders we have a larger proportion of iron surface to the volume of steam used per stroke and per minute.

We will refer more at length to compounding of steam-cylinders, after having determined the value of C in a number of cases.

(81.) Computation of Condensation.—Bearing in mind these many qualifying conditions surrounding the actual use of steam in the engine, let us take up and compare the condensation in the case of several different engines. (Table, page 206.)

The experiments of J. W. Hill were made to decide the relative economy of three types of engines, built by rival makers, and were published in pamphlet form, after the tests, without award, as the results were very close.

The experiments of Messrs. Gately and Kletzsch were made with a view to determining the condensation of steam-cylinders. Their general course was directed by Professor Thurston, to whom the writer had previously communicated his views and published writings on initial condensation in steam-cylinders.

18

Table of Calculations of Condensation of Steam by Surface of One Square Foot of Cast Iron for One Degree Fahrenheit and One Minute.

Reference Number of Experiment.	Duration of Experiment. (h. m.)	Stroke of Cylinder in Feet plus Clearance. (s)	Diameter of Cylinder in Feet. (d)	Reciprocal of True Number of Expansions. (e)	Number of Strokes per Minute. (N)	Quality of Steam used (British Units). Thermal Value.	Absolute Pressure in Pounds per Square Inch at Cut-off. (P_b)	Maximum Compression Pressure, Pounds per Square Inch. (B_m)	Absolute Steam-Pressure of Exhaust at Mid-Stroke, Pounds per Square Inch. (B)	Temperature of Cylinder at Cut-off (Fahr.). (T_b)	Temperature of Cylinder for Maximum Compression. (T_c)	Temperature of Cylinder for Exhaust. (T_e)	Number of Degrees of Superheating (Fahr.). (T_h)	Temperature of Superheated Steam for P_b. (T_s)	Specific Volume of Saturated Steam at Cut-off. (S)	Specific Volume of Superheated Steam at Cut-off. (S_s)	Ratio of Actual to Indicated Steam at Cut-off. (r)	Condensation Constant In Lbs. of Steam. (C)	Condensation Constant In British Units of Heat. (C)	Actual Steam per Indicated H.P. per Hour (Weighed). (W_a)	Calculated Steam per Indicated H.P. per Hour at Cut-off. (W)	Value of Y from preceding Value of e. (Y)
Clothed. (18″ × 48″) Harris-Corliss Engine. Experiments of J. W. Hill, Cincinnati, 1880.																						
(1)	10.00	4.077	1.502	0.135	151.66	(?)1315.86	101.49	26.59	3.35	328.65	244.81	145.63	203.36	532.01	264.35	386.67	1.634	0.02490		19.364	11.848	0.32
(2)	10.00	4.077	1.502	0.153	151.62	(?)1255.74	100.43	60.62	14.94	327.88	293.19	212.80			266.87		1.513	0.03734		23.907	15.799	0.341
Clothed. (18″ × 48″) Wheelock Engine. Experiments of J. W. Hill, Cincinnati, 1880.																						
(3)	10.00	4.094	1.522	0.151	148.94	(?)1301.65	92.15	28.10	4.71	321.7	296.50	159.64			288.50		1.541	0.02312		13.5	12.638	0.338
(4)	10.00	4.094	1.522	0.189	152.14	(?)1313.11	91.36	58.71	15.45	321.10	291.11	214.5			290.80		1.423	0.03610		24.6	17.514	0.371
Clothed. (18″ × 48″) Reynolds-Corliss Engine. Experiments of J. W. Hill, Cincinnati, 1880.																						
(5)	10.00	4.106	1.502	0.147	150.77	(?)1243.84	100.93	13.64	4.45	328.25	208.24	157.24			265.7		1.567	0.03		20.618	13.157	0.334
(6)	9.00	4.106	1.502	0.181	150.66	(?)1211.30	99.37	49.32	15.56	327.11	279.99	214.85			269.2		1.441	0.04		25.945	18.009	0.372

Average, 0.02932.   25.934 British Units.

Ref. No.	Duration (h. m.)	Stroke s	Diam. d	Recip. e	No. Strokes N	Quality / Thermal Value	P_b	B_m	B	T_b	T_c	T_e	T_h	T_s	S	S_s	r	C (Lbs.)	C (Br. Units)
(7)	1.40	3.5	1.5	0.589	136.52	Dry Sat.	61.54		4.22	294.2		155.2					1.294	0.01576	14.29
(8)	2.00	3.5	1.5	0.443	135.9	"	68.34		3.91	301.14		152.15					1.371	0.01929	18.01
(9)	1.55	3.5	1.5	0.330	134.64	"	62.10		4.48	294.80		157.56					1.512	0.01909	17.31
(10)	2.00	3.5	1.5	0.131	137.9	"	49.11		3.65	279.84		149.4					2.003	0.01652	15.15
(11)	2.00	3.5	1.5	0.208	138.03	"	78.80		3.24	311.02		144.58			342.2		1.544	0.01544	13.82
(12)	1.45	3.5	1.5	0.206	84	"	66.89		3.83	299.69		068			398.1		(?)1.917	(83)	(?)
(13)	2.00	3.5	1.5	0.94	143.46	"	53.21		3.24	284.95		44.58			492.1		1.583	0.01416	12.94
(14)	2.00	3.5	1.5	0.210	137.82	"	39.83		3.61	266.74		149.02			62.9		1.707	0.01528	14.16
(15)	2.00	3.5	1.5	0.242	135.85	"	26.74		3.46	244.26		147.21			936.6		1.700	0.01372	12.94
(16)	3.00	3.5	1.5	0.412	135.96	"	65.36		14.70	298.16		212.			407.5		(?)1.122	0	94 (?)
(17)	2.30	3.5	1.5	0.420	137.14	"	50.42		14.82	281.30		212.4			64.5		(?)1.307	08	98 (?)
(18)	3.00	3.5	1.5	0.401	135.02	"	10.52		14.88	267.87		212.6			620.8		30	0.01416	13.11
(19)	3.00	3.5	1.5	0.466	133.04	"	28.40		14.84	247.56		212.49			86.6		(?)1.376	00	90 (?)
(20)	1.30	3.5	1.5	938	125.95	"	27.38		3.15	245.56		146.31			916.2		82	0.01404	12.23
(21)	2.00	3.5	1.5	0.961	100.60	"	8.35		3.86	247.46		151.57			886.8		1.403	08	14.36
(22)	1.45	3.5	1.5	0.981	67.48	"	28.53		4.96	247.81		161.94			881.6		00	90	13.62
(23)	10.00	1.722	0.833	0.208	402.22	Dry Sat.	95.09	92.32	15.86	324.		215.8		280.3			1.526	0.04414	

(18″ ×) Harris-Corliss Engine. Clothed.

cits of Messrs. Gately and Kletzsch, Sandy Hook, 24. — Min. 14.

Experiments of Assistant-Engineer H. W. Spangler, U.S.N., Philadelphia Electrical E Clothed. (10′ × 20″) Buckeye Engine, Slide-Valve Riding Cut-off.

14.33. Average of Gately and Kletzsch Experiments.

For Comparison.
Results of Hill's Experiments, Compression Neglected.

	r	C	C
(1)	1.5730	0.01808	
(2)	1.4046	0.02025	
(3)	1.4695	0.01511	
(4)	1.3102	0.01843	
(5)	1.5351	0.01762	
(6)	1.3372	0.01933	

Average 0.01814 = 16.045 British Units for a Pressure of 97 Lbs.

{ 28.579 Lbs. per H.P. Corrected for moisture.

The experiments of the International Electrical Exhibition of 1884 were conducted, under a code drawn up by the writer and approved and adopted by the Franklin Institute, by H. W. Spangler, Assistant-Engineer, U.S.N.

In the cases in the preceding table the piston was known to be tight and a trifling leakage to occur at the valves.

Unless the piston and valves are proved to be tight, experiments upon steam-engines have no scientific value.

For this reason these results have not that absolute accuracy so desirable in all scientific work; but they represent the best results attained by our mechanics skilled in engine-building, and for that reason may be of greater practical value than results derived from more accurate work under conditions not realizable in practice.

(82.) **Probable Errors of the Table** (pages 206, 207). —In Mr. Hill's experiments, his extreme care in computation renders the possibility of error in the value of the constant of condensation very slight. The difference observable in each engine, condensing and non-condensing, is due to the slight leakage at the valves, to the probable errors of the indicators used, and to the practical impossibility of closely measuring the exact points and pressures on the indicator diagrams, as well as to the *short time of exposure to compression temperatures.*

Mr. Hill writes as follows:

"CINCINNATI, February 2, 1884.

" DEAR SIR,—The calorimeter used in the tests for quality of steam at the Miller's Exhibition was of the continuous kind, carefully made, and in charge of my principal assistant.

" I accepted his notes and data as correct at the time, but from more recent experiments am inclined to doubt any result from a calorimeter of this kind, owing to the fact of large variations in the temperature of overflow (known to subsist by other circumstances) not recorded by the ordinary mercurial thermometer.

"I now use a simple arrangement of tub, scale, and hand thermometer, and, while this method is liable to error, the error is not sufficient to lead you astray.

"For the purpose of comparison I regard the calorimeter data of the Miller's Exhibition trials as correct, but cannot endorse it for absolute results.

* * * * * * * * *

"I know the Harris piston was tight, from special test.

"*Page* 74. There is no doubt all the engines suffered from leaks into and out of cylinder through steam- and exhaust-valves."

* * * * * * * * *

There is no doubt in the mind of the writer that super-heated steam did not reach these engines. It would appear from the results attained that all of the engines were tight as to pistons. Messrs. Gately and Kletzsch state that the engine upon which they experimented was tight in valves and piston.

When compression is neglected, as is done for the purpose of comparing Mr. Hill's work with that of Messrs. Gately and Kletzsch, it will be seen that the variation from the mean value of $C = .01814$ pound of steam is remarkably small. It would seem probable, however, that the true value of C is greater than 16.045 British units, which, however, will prove the most useful constant for general work in which compression is not closely regarded.

Especially where the clearance is very small the exhaust closure is very late, and the time of exposure of the piston- and cylinder-heads so very short as would probably prevent the temperature due to compression from acting with its full effect. (See expression given.) We should neglect the temperatures due to compression, but not the steam saved by it.

The experiments on the condensing engines 1, 3, and 5 will probably give the most accurate values of C. All of these engines were built by our best makers for the special purpose of a competitive test, and, most fortunately, the elaborate record of them was made by a thorough and experienced engineer.

The experiments of Messrs. Gately and Kletzsch, although marked by less care and precision, supplement Mr. Hill's work, by enabling us to prove the law of condensation to be true under all conditions as to speed, initial exhaust, and back pressures and points of cut-off. Their short duration gives them less value and renders them less reliable as to absolute results. The clearance should have been experimentally determined. Diagrams should have been published in conjunction with the record, and the amount of compression carefully noted, as it doubtless has an important effect upon the initial condensation of steam.

These concordant experiments were made with pressures varying from 102 to 27 pounds at cut-off, with speeds varying from 152 to 67 strokes per minute, with the true cut-off varying from .131 to .981 of the stroke, with and without condenser, and by different experimenters upon different engines.

The condensation for the Buckeye engine is roughly computed to show the injurious effect of a slide-valve bathed in exhaust-steam. Although this engine was not tested for tightness, the deservedly high reputation of its makers would render improbable any serious loss by leakage.

(83.) **The Influence of the Condenser.**—The condenser adds about 10 pounds to the mean effective pressure upon the piston; it cools the exhaust about 60° Fahrenheit; it demands power for pumping; and it may, when the engine is too lightly loaded, increase the actual consumption of steam per horse-power per hour. The sudden increase in the fall of temperature of steam in the last three or four pounds of pressure above a vacuum has taught makers, by practical results, not to carry their vacuums too far down.

From the indicator-card can be calculated the amount of steam from the boiler required to do the work and meet the demands of condensation. Equations (219) and (223 *a*.)

(84.) **The Jet-Condenser.**—The jet-condenser is the cheapest and most efficient, and should be used whenever the quality of the water will permit.

Whatever temperature for the mixture of condensing water and steam is required will give the weight of condensing water from the following equation :

$$W_a H_a + W_c H_c = (W_a + W_c) T_1,$$

in which

W_a = actual steam per indicated *HP* per hour in pounds.
H_a = British units of heat in one pound of steam at boiler.
W_c = condensing water in pounds per hour per *HP.*
H_c = its heat units per pound.
T_1 = the approximate temperature of water flowing from the condenser (Fahr.).

(85.) **The Surface-Condenser.**—The surface-condenser should only be used when it is necessary to separate the condensing water from the boiler feed-water, and, as a matter of convenience, it should have separate circulating and exhaust-pumps. In most cases, although it is a proceeding less economical of coal, it will be found to have many advantages. It is usual among designers to allow from two to four square feet of condensing surface per indicated horse-power.

Isherwood states (Shock's *Steam-Boilers*, page 38) that the conductivity of metal plates is independent of their thickness, and that for a difference of temperature of one degree Fahrenheit one square foot will transmit the following numbers of heat units per hour :

Copper.................... 642.543 British units.
Brass..................... 556.832   "   "
Wrought iron........... 373.625   "   "
Cast iron................. 315.741   "   "

It is not safe to reckon more than 10 to 20 per cent. of the above values in actual condensers. The steam carries with it the cylinder lubricants, which foul the surfaces of the tubes, and the condensed steam collects on the tubes, preventing their rapid action, while the high vacuum may materially degrade the efficiency of the condensing surfaces.

The steam released from the cylinder at the terminal pressure of expansion still further expands into a body of steam in the condenser, and then condenses.

The circulating condensing water rises in temperature from 20° to 40° Fahr. The larger the condensing surface per indicated horse-power, the higher the allowable rise of temperature. To be on the safe side, the estimated transmissive power of the condensing surfaces should not exceed 20 per cent. or less of Isherwood's values, and the temperature of the steam in the condenser should be estimated from the vacuum-pressure.

Let C_1 = constant of transmission.
" T_v = temperature of vacuum.
" T_k = mean temperature of circulating water.
" H_a = total heat of steam at boiler.
" A = condensing surface.

Then

$$A = \frac{W_a(HP)\,(H_a - T_k)}{10 \text{ to } 20\% \text{ of } C_1(T_v - T_k)}.$$

Let $T_k = \dfrac{T_1 + T_2}{2}$.

" T_1 = temperature of entering circulating water.

" T_2 = temperature of departing circulating water.

" W_c = weight of circulating water in pounds.

" T_v = heat units of condensed water.

Then we have

$$W_c = \frac{\{(HP)\,(H_a - T_v)\}}{T_2 - T_1}\,W_a.$$

Both of these formulæ are approximations, giving sufficiently close values for an estimate of cost of making and of power required by pumps. Much more frequently than is believed, additional expense for power is the result of adding a condenser.

(86.) **The Law of Re-evaporation after Cut-off.—** Whenever we examine the indicator-cards correctly taken from engines with practically tight valves and piston, not embarrassed by sluggish valve-motions, we find the curve of expansion to follow very closely the equilateral hyperbola. We also find that the quantity of steam present as vapor is a minimum at the point of cut-off, and that it steadily increases in volume as the expansion goes on, reaching a maximum at the point of release.

The added steam appearing at the point of release should come from the condensation at the cut-off, which condensation should be decreased in conformity with the law of condensation already stated.

Let T_f = temperature of pressure at release.

" Y^1 = value of curve of condensation for e^1 (at release).

We can then write from equation (221),

$$\frac{\text{Condensation at cut-off}}{\text{Condensation at release}} = \frac{(T_b - T_e)\,(d + Ys)}{(T_f - T_e)\,(d + Y^1s)}.$$

Mr. Hill computed the steam from his diagrams at the

point of release, the writer for the point of cut-off from the same diagrams.

A comparison will afford additional verification of the law of condensation, and show the law of re-evaporation to be identical. Writing this in its most complete form, we have

$$(r^1 - 1) = (r - 1) \frac{(T_b - T_c)d + (T_f - T_e)Y^1_s}{(T_b - T_c)d + (T_b - T_e)Y_s}.$$

The following table gives the results computed by this formula, and also the results of Mr. Hill's independent calculations and measurement:

Reference Number of Experiment.	Fraction of True Stroke for Release.	Value deduced from Preceding Column.	Absolute Pressure at Release (Lbs. per sq. in.).	Temperature at Release (Degrees Fahrenheit).	Calculated Steam per H.P. per Hour at Release (Hill's Figures.)	Ratio of Actual to Indicated Steam at Release.	Percentage of Condensation at Release.					
	e^1	Y^1	P_f	T_f	W^1	r^1	r^1-1					
1	0.94	.967	14.57	211.8	13.755	1.408						
2	0.97	.983	17.04	219.5	18.013	1.327						
3	0.96	.978	14.04	209.6	13.915	1.400						
4	0.97	.983	17.46	220.8	19.674	1.267		From Preceding Table.				
5	0.98	.989	15.16	213.6	14.886	1.385						
6	0.98	.989	17.41	220.6	18.896	1.373		$r-1$	T_b-T_c	T_f-T_e	T_b-T_e	Y
1, 3, and 5 averages		.978	14.44	211.6	....	1.395	.355	.581	93.	57.4	172.	.331
2, 4, and 6 averages		.985	....	220.3	....	1.323	.238	.459	40.6	6.3	114.7	.361

The calculated value falls short of the observed value by 4 per cent., or about ½ pound of condensation per horse-power per hour for the engines with condenser. For the engines non-condensing the calculated value falls short 8 per cent. of the indicated steam per horse-power per hour, or about 1½ pounds.

A certain amount of this condensation is due to heat converted into work. This difference is in the expected direction, for the engines were none of them steam-jacketed, and, being subject to loss from radiation, some additional lack would naturally be anticipated. The undoubted leaks into and out of valves would appear to affect the diagrams of the engines non-condensing most seriously.

It would also appear that the intensity of the pressure of the compressed steam facilitated its condensation, otherwise the results calculated with compression would not agree better with the facts than when calculated without compression, as can be seen by trial.

It is of particular importance to show the practical identity of the laws of condensation and re-evaporation as a preliminary to the discussion of compounded cylinders.

(87.) **The Influence of Compounding Cylinders.**— We have carefully followed the action of the steam while passing through one cylinder. Let us follow it through the two cylinders. The steam entering the non-condensing cylinder suffers initial condensation, in some instances quite copious, but not so great as if the non-condensing cylinder had the temperature of the exhaust, T_e.

The steam being cut off in the non-condensing cylinder, re-evaporation begins, the expansion-line being held closely to an equilateral hyperbola.

This re-evaporation is, however, far from being complete, and at the end of the stroke communication is opened to the condensing cylinder. At this instant a relatively enormous initial condensation occurs, because of the great surface of condensing piston- and cylinder-head presented at the temperature of exhaust; but this condensation is met at once by the equally active re-evaporation which simultaneously occurs from the whole interior of the non-condensing cylinder, the result being the transferring of the condensation from the surface of the non-condensing cylinder to the surface of the condensing cylinder until the temperatures are equalized.

After the violence of this first transfer of condensation has abated, the re-evaporation from the interior of both cylinders occurs with sufficient celerity to hold the expansion-curve closely to an equilateral hyperbola.

If but two cylinders are used, the condensing cylinder is now opened to exhaust, and the re-evaporated and vaporous steam enter the condenser, carrying much more heat than would appear from a calculation of the thermal value of the vaporous steam present at the end of expansion.

Thus we see that at the present day the compound engine owes its possible greater efficiency to the physical attributes of iron rather than to the properties of steam, and that with the use of non-conducting materials the necessity of compounding cylinders will vanish.

When cranks of compounded cylinders are placed at right angles, and the steam is cut off from the condensing cylinder at an early point in the stroke, certainly earlier than one-half stroke, it will be found that a considerable increase in the economy will occur with a small receiver, although this arrangement will cause trouble in equalizing the power of the cylinders, because of the increase of back pressure in the non-condensing cylinder until its piston reaches mid-stroke.

This economy arises from the fact that the transferrence of condensation from the non-condensing cylinder to the condensing cylinder is greatly facilitated by an increased difference in temperatures of the non-condensing cylinder and receiver and of the condensing cylinder.

The particularly injurious effect of a double admission of steam to the condensing cylinder when cranks are at right angles and the cut-off of the condensing cylinder is later than one-half stroke, arises from the fact that this re-evaporation from the iron surfaces is temporarily stopped by the entrance of steam of a higher pressure and temperature from the non-condensing cylinder.

The advantage of the compound engine must lie in its lesser condensation alone, other things being equal; and this diminution of condensation must compensate for the increased quantity of machinery demanded before we begin to consider its superiority.

This point must be considered experimentally by a careful determination of the ratio of the actual to the indicated steam and heat.

For the purpose of gaining a clear, general idea, let us assume a perfect gas, expanding according to Mariotte's law, fed to a pair of compounded cylinders at a given initial pressure, P_b, and exhausted against a back pressure, B, outside of the cylinder. While these assumptions will not perfectly fulfil the conditions of steam, the results obtained will serve as a guide in the use of steam, and by proper modifications can be applied to the steam-engine itself. When steam is used, the high initial temperature of the steam is communicated to the walls previously at or above the temperature of exhaust by means of the condensation of the steam which results in the water ready for re-evaporation at the instant of any diminution of pressure. That part of the cylinder-walls subjected to initial steam being hotter than the expanded steam gives the steam this heat very readily, which goes for the twofold purpose of reereating steam and of warming up to the temperature of the steam the gradually uncovered walls of the cylinders, which are at the temperature of the exhaust or perhaps above it.

These exchanges go on with a celerity not easily apprehended without thoughtful consideration of the great weight of iron in the steam-cylinder, and its conductivity for heat as compared wi h the relatively exceedingly small weight of steam and water at the temperature of evaporation, which enter and leave the cylinder at each stroke.

Theories to the contrary notwithstanding, it would seem as if within the limits of economic expansion an equilateral hyperbola represents, with quite as great approximation as any other curve, the pressures of expanding steam in an iron cylinder steam-jacketed or clothed.

FIG. 40.

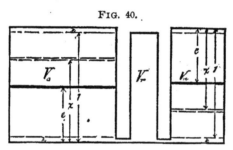

Taking, then, a perfect gas as our starting-point, and assuming as the simplest case two cylinders with cranks 180 degrees apart,

Let V_n = the true volume of the non-condensing cylinder (including one of its clearances, k) up to its valve-face.

" V_r = the volume of the connecting channels (and receiver, if there is one) from valve-face to valve-face.

" V_c = the true volume of the condensing cylinder (including one of its clearances, k_1) up to its valve-face.

" e = the true cut-off of the non-condensing cylinder = the reciprocal of the true expansion in it.

" P_b = the initial pressure of the non-condensing cylinder, pounds per square inch abs.

" e_1 = the true cut-off of the condensing cylinder = the reciprocal of the expansion in it after cut-off.

" B = the back pressure of the condensing cylinder, pounds per square inch abs.

19

After a compound engine has attained its regular work, it has attained such a pressure, P_r, in the receiver (the word receiver will be used to comprise all pipes, steam-chest, etc., that may be between the two cylinders, whether there be a specially-designed receiver or not) as will enable the condensing cylinder to void the same weight of vapor at each stroke as is received by the non-condensing cylinder at each stroke, and we can therefore write the equation

$$e\,P_b\,V_n = e_1 P_r\,V_c$$

$$P_r = \frac{e P_b V_n}{e_1 V_c} = \frac{e P_b}{R e_1}.$$

The pressure P_r is that occurring at the exact moment when the port to the non-condensing cylinder is closed, and when the piston-head of V_c is at a distance e_1 from the beginning of its volume. (See Fig. 40.)

The absolute mean pressure pressing forward upon the non-condensing piston-head can be written

$$e P_b \left[1 + \text{nat. log } \frac{1}{e} \right] - P_b k.$$

At the moment of the closing of the steam-port V_c the pressure P_r exists in all three divisions, pressing forward in V_c upon the piston, pressing backward in V_n upon the piston.

The backward pressure upon the piston of V_n can now be calculated at any point from the beginning of stroke to e_1.

$$P_r [(1 - e_1 + k) V_n + V_r + e_1 V_c] = P_x [(1 - x + k) V_n + V_r + x V_c].$$

Therefore the mean back pressure absolute reduced to a full stroke upon the non-condensing piston, while the two pistons proceed in opposite directions through a fraction of the volume $(e_1 - k)$, is

$$\frac{P_r[(1 - e_1 + k) V_n + V_r + e_1 V_c]}{(V_c - V_n)} \text{ nat. log } \frac{(1 + k)V_n + V_r + (V_c - V_n)e_1}{(1 + k) V_n + V_r + (V_c - V_n)k}.$$

This is also the mean forward pressure, absolute, upon the piston of the condensing cylinder.

When, now, the port of the condensing cylinder is closed, the two pressures part company, the back pressure in the non-condensing cylinder rising first by compression into the receiver and itself, and when the non-condensing exhaust is closed rising still more rapidly by compression into the clearance of V_n, and the vapor in the condensing cylinder expanding and its pressure falling.

Let us consider first the back pressure in the non-condensing cylinder, while the piston moves through the fraction of the stroke $(1-e_1)$. We can write the following equation:

$$P_r[(1-e_1+k)\,V_n + V_r] = P_x[(1-x+k)\,V_n + V_r].$$

Therefore we have the mean absolute back pressure, when the non-condensing exhaust is not closed till the end of the stroke,

$$\frac{P_r[(1-e_1+k)\,V_n + V_r]}{V_n} \text{ nat. log } \frac{V_r + V_n(1+k-e_1)}{V_r + V_n k}.$$

Secondly, the forward expansion pressure in the condensing cylinder, after its port to the receiver is closed. We can write the following equation:

$$P_r[e_1 V_c] = P_x x\,V_c.$$

Therefore its absolute mean pressure is

$$e_1 P_r \text{ nat. log } \frac{1}{e_1}.$$

Finally, if compression is used in the condensing cylinder, the point at which compression begins being the fraction b of the volume of the condensing cylinder, we have

$$B\left[1 - b\left(1 - \text{nat. log } \frac{b}{k_1}\right)\right].$$

We can now write the expressions for the mean effective pressure in each cylinder.

For the non-condensing cylinder we have, as the expression for the work done by it in one stroke,

$$V_n \left\{ eP_b \left[1 + \text{nat. log} \frac{1}{e} \right] - P_b k - \frac{P_r [(1 - e_1 + k) V_n + V_r + e_1 V_c]}{(V_c - V_n)} \right.$$

$$\text{nat. log} \frac{(1+k) V_n + V_r + (V_c - V_n) e_1}{(1+k) V_n + V_r + (V_c - V_n)k} - \frac{P_r[(1 - e_1 + k) V_n + V_r}{V_n}$$

$$\left. \text{nat. log} \frac{V_r + V_n(1 + k - e_1)}{V_r + V_n k} \right\}. \qquad (227\,a)$$

The expression for the work done by the condensing cylinder during one stroke is, if we assume $k = k_1$,

$$V_c \left\{ \frac{P_r[(1 - e_1 + k) V_n + V_r + e_1 V_c]}{V_c - V_n} \text{ nat. log} \right.$$

$$\frac{(1+k) V_n + V_r + (V_c - V_n) e_1}{(1+k) V_n + V_r + (V_c - V_n) k_1} + e_1 P_r \text{ nat. log} \frac{1}{e_1}$$

$$\left. - B \left[1 - b \left(1 - \text{nat. log} \frac{b}{k_1} \right) \right] \right\}. \qquad (227\,b)$$

Assume $V_r = 0$, $k = 0$, $k_1 = 0$, $B = 0$, and therefore $e_1 = 1$.

Equate equations (227 a) and (227 b) under these assumptious, we have then the following criterion in order to equalize the power of the cylinders:

$$\frac{2R}{R-1} = \frac{\log 2.7183\, E}{\log R}. \qquad (227\,c)$$

There is no fall in pressure from non-condensing to condensing cylinder when communication is opened between them.

If we do not assume $B = 0$, we have

$$\text{nat. log } E = \frac{2R}{R-1} \text{ nat. log } R - \left[1 + E \frac{B}{P_b} \right];$$

19*

or, in common logarithms,

$$\log. E = \frac{2R}{R-1} \log. R - \frac{\left[1 + E\frac{B}{P_b}\right]}{2.3026}.$$

Approximate values of R, E, and e under these conditions will be found tabulated below, the term $E\frac{B}{P_b}$ being neglected as being very small, as it usually is, condensing.

(88.) The Relation of Cylinder Ratio to Ultimate Expansion.

Table of Ratios of Cylinders and of Points of Cut-off in Non-Condensing Cylinder for Ultimate Expansions of Steam.

Criterion $\dfrac{2R \log R}{R-1} = \log 2.7183\ E$ for equal powers.

$E = R^2 + R$ for equal initial thrusts.

Ratio of Cylinder Volumes.	Ultimate Expansion of Steam.	Point of Cut-off in Non-Condensing Cylinder.	REMARKS.	For Equal Initial Thrusts.
R	E	$e = \dfrac{R}{E}$		E
$1\frac{1}{4}$	3.426	0.37	These computations are made for general guidance only, nearly all the assumptions made being impossible of exact realization.	2.812
$1\frac{1}{2}$	4.190	0.35		3.750
$1\frac{3}{4}$	5.015	0.34		4.812
2	5.886	0.34	It is assumed that a perfect gas is used expanding isothermally, that there is no back pressure on the condensing cylinder piston, and that there are no clearances or receiver, and further that there is no cut-off at all for the condensing cylinder, that only being demanded to provide for receivers and clearances in actual practice, and that the cranks are together, or 180 degrees apart.	6.000
$2\frac{1}{4}$	6.813	0.33		7.312
$2\frac{1}{2}$	7.801	0.32		8.750
$2\frac{3}{4}$	8.840	0.31		10.312
3	9.933	0.30		12.000
$3\frac{1}{4}$	10.961	0.30		13.812
$3\frac{1}{2}$	12.277	0.28		15.750
$3\frac{3}{4}$	13.580	0.28		17.812
4	14.832	0.27		20.000

If we require equal initial piston thrusts at the commencement of each stroke, we have

$$P_b - e P_b = [e P_b - B] R \text{ or } E = \frac{R^2 + R}{1 + R\dfrac{B}{P_b}}.$$

It is useless to carry this table further. Enough has been done to show the error of exaggeration of ratio into which designers have fallen, when it is necessary to equalize the power of cylinders and to avoid an intermediate drop for the sake of economy.

Indeed, in all engines it will be found that economy of steam as well as smoothness of action demands that no sudden changes of pressure shall be permitted, and therefore it will be found advantageous in single cylinder engines where clearance cannot be indefinitely reduced to use enough compression to bring the back pressure up to the initial pressure, and so as not to permit an *explosion* in the cylinder at the beginning of each stroke.

The less the clearance, the less the compression required for this purpose, and consequently the less the power of the engine is absorbed in fulfilling this condition.

(89.) **The Influence of the Receiver and Clearances.**—If the receiver be considered, it is obvious that the only result of increasing its proportions is to decrease the mean pressure of the steam against both pistons up to the point of cut-off e_1 of the condensing cylinder.

This will decrease the power of the condensing cylinder and increase the power of the non-condensing cylinder up to the point of cut-off e_1.

After steam is cut off in the non-condensing cylinder the back pressure is not so rapidly raised with a larger receiver, which results in a further increase of the power of the non-condensing cylinder.

After steam is cut off in the condensing cylinder, its power is in no wise affected by the size of the receiver, as the pressure $P_r = \dfrac{eP_b}{e_1 R}$ depends on the initial pressure, the ratio of the volumes of the two cylinders, and their respective points of cut-off.

This pressure P_r occurs when the steam is just being cut off from the condensing cylinder; it can also be written

$$P_r = \frac{P_b}{Ee_1}.$$

We observe that when the capacity of the receiver is assumed very great, the back pressure line of the non-condensing cylinder comes very near being a straight line; and further, if we make e_1 equal to unity, the forward pressure line of the condensing cylinder comes near a straight line, and $P_r = \dfrac{eP_b}{R}$.

The ultimate expansion by pressures is

$$E = \frac{P_b}{e_1 P_r} = \frac{V_c}{eV_n}.$$

That is to say, it is theoretically quite independent of the volume of the receiver, as also of the point of cut-off in the condensing cylinder.

If we fix the ultimate expansion E of the steam and the point of cut-off in the non-condensing cylinder, we at once determine the ratio R of the volumes of the two cylinders:

$$R = Ee = \frac{V_c}{V_n}.$$

If we wish, having assumed a certain ultimate expansion E and ratio of volumes of cylinder R, to determine at what point the condensing cylinder must cut-off in order to render the work for the steam or gas a maximum, we must so arrange that the terminal pressure eP_b shall equal the pressure of the steam in the receiver at the moment of its admission: this would require

$$\frac{P_b}{Ee_1} = \frac{eP_b \, (kV_n + V_r)}{(1 - e_1 + k) \, V_n + V_r}.$$

With a fixed volume of receiver we have

$$e_1 = \frac{(1+k) \, V_n + V_r}{V_n + Ee \, [kV_c + V_r]}. \tag{228}$$

From the equation for the compression pressures in the non-condensing cylinder we can write the value of x = the fraction of the true volume at which its exhaust-port must be closed, so that the pressure in the receiver shall not rise above the final pressure eP_b.

$$x = 1 + k + \frac{V_r}{V_n} - \frac{1}{Re_1} \left\{ (1 - e_1 + k) + \frac{V_r}{V_n} \right\}.$$

For the mean absolute back pressure while the non-condensing piston passes from e_1 to x we have

$$P_r \frac{(1 - e_1 + k) \, V_n + V_r}{V_n} \text{ nat. log } \frac{V_r + V_n(1 + k - e_1)}{V_r + V_n (1 + k - x)}.$$

We have also the following expression for the absolute mean back pressure on the non-condensing piston while compressing from x to k:

$$eP_b (1 - x + k) \text{ nat. log } \frac{x}{k}.$$

That is to say, the presence of a receiver is always productive of a loss where two cylinders are worked together, regardless of its influence on the cut-off of the condensing cylinder, *and it is entirely unnecessary when cranks are together or 180 degrees apart.*

Unfortunately, we cannot suppress the clearances altogether, and therefore from the above we have, assuming $V_r = 0$,

$$e_1 = \frac{(1+k) \, V_n}{V_n + Rk_1 \, V_c}.$$

FIG. 40.

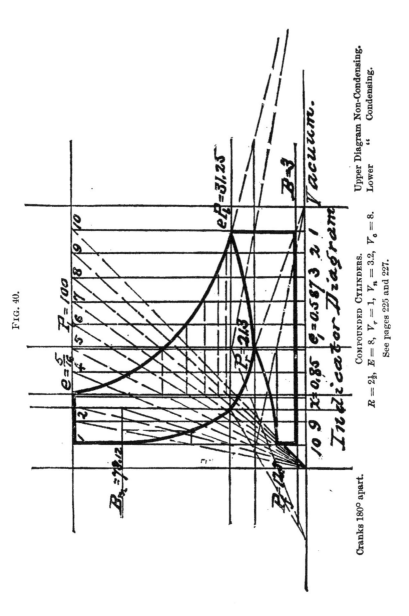

Upper Diagram Non-Condensing.
Lower   "   Condensing.

COMPOUNDED CYLINDERS.

$R = 2\tfrac{1}{2}, \; E = 8, \; V_r = 1, \; V_n = 3.2, \; V_o = 8.$

See pages 225 and 227.

Cranks 180° apart.

V_r will represent the volume of the connecting-pipes, and require its use in actual practice.

Example.—Let $P_b = 100$ pounds per square inch (absolute).

" $E = 8$.   Then $R = 2\frac{1}{2}$.   Let $B = 3$ pounds abs.

" $V_o = 8$.   Then $V_n = 3.20$, and $e = \dfrac{5}{16}$.

" $V_r = 1$.   Let $k = k_1 = b = 10\%$.

$$e_1 = \frac{3.52 + 1}{3.20 + 2\frac{1}{2}(1 + .8)} = \frac{4.52}{7.7} = 0.587. \quad \text{See equation (228).}$$

$$P_r = \frac{eP_b}{Re_1} = \frac{31.25}{2\frac{1}{2} \times 0.587} = 21.3 \text{ pounds.}$$

$$\left.\begin{array}{l}\text{Point of exhaust-closure of} \\ \text{non-condensing cylinder}\end{array}\right\} x = 0.85.$$

Power of non-condensing cylinder $= 69.03 - [12.49 + 6.75 + 7.16] = 42.63.$ (227 *a*)
Power of condensing cylinder $= 2\frac{1}{2}[12.49 + 6.66 - 2.7] = 41.12.$ (227 *b*)

Had there been no back pressure, B, and no clearances or receiver, the cylinders would have balanced. Another approximation will be sufficient, provided it is not deemed that losses in the non-condensing cylinder and passages will reduce its power sufficiently when steam is used.

(90.) The Horse-Power of Compounded Cylinders. —When these points have been covered, we must determine the size of the non-condensing cylinder for the required horse-power.

If, now, we assume the case of compounded cylinders, with cranks together or 180 degrees apart, no receiver, no drop in the expansion, no cut-off on the condensing cylinder, and no clearances or compression, we have, from addition of the equations for the work of the two cylinders (227 *a*) and (227 *b*),

$$PV = V_n \left\{ eP_b \left[1 + \text{nat. log} \frac{R}{e} \right] - RB \right\},$$

or, in terms of the volume of the condensing cylinder and of the ultimate expansion, since $\dfrac{R}{E} = e$,

$$PV = V_e \left\{ \frac{P_b}{E} \left[1 + \text{nat. log } E \right] - B \right\}.$$

That is to say, mathematically it can be shown that, the same measure of expansion being used in both cases, the power of the condensing cylinder alone is equal to the combined powers of the two cylinders of a compound engine, and if we neglect initial condensation the steam economy is the same.

The mean effective pressure, from the above equation, of the condensing cylinder into its volume gives the work in one stroke $= PLA$, and the horse-power is

$$(HP) = \frac{pLAN}{33000}.$$

The rest of the proportioning follows from what has heretofore been proved.

(91.) **The Influence of Cranks at Right Angles.**— Cranks of compound engines are placed at angles less than 180°, for convenience of handling engines which have no fly-wheel, or which have to be stopped and started, or reversed: this procedure requires the presence of a receiver or compression space, and converts the distances of the pistons into trigonometric functions relatively to each other, but in no wise alters the principles involved.

When compound engines must be frequently started or reversed, it is important that they be so arranged as to avoid getting on their centres; this does not apply to pumping engines, or indeed to the majority of stationary engines, but does apply with a great deal of force to marine engines.

Some designers have endeavored to obviate this difficulty

by placing the cranks 160 degrees apart, thus enabling a very small dead space between the cylinders, and obviating trouble with the engine on its centre.

This case differs so little from the case already considered with the cranks together or 180 degrees apart, that it requires but one precaution on the part of the designer:

The non-condensing cylinder should exhaust *before* the condensing cylinder takes steam, not after, as that would cause a sudden rise in the pressure of the condensing cylinder, with the attendant loss.

A good deal of weight is laid on equalizing stresses by placing the cranks 90 degrees apart; but, as a very large number of engines of the first type have been successfully designed, there is no reason to believe it impossible in the future to use engines with cranks 180 degrees apart.

It is quite possible to give sufficiently large clearance-spaces in the case of engines with cranks at right angles to fulfil to a large extent the functions of a receiver, but it is more economical to reduce the clearances of the steam-cylinders as much as possible and to provide a receiver of proper size, since this will avoid sudden changes in pressure and the consequent loss.

In what follows we will neglect the variation of the piston's positions due to the angular position of the connecting-rods.

Assuming the engine to have obtained its regular movement, we will have $P_r = \dfrac{eP_b}{e_1 R}$.

The sequence of exhaust from the non-condensing cylinder, and of cut-off of the condensing cylinder, renders necessary the presence of a receiver or its equivalent when cranks are at right angles. The exhaust from the non-condensing cylinder occurs just as the piston of the condensing cylinder reaches mid-stroke, and so, in order to avoid a sudden

20

change of pressure and of the progress of expansion in the condensing cylinder, it is necessary that its cut-off be earlier than one-half stroke.

The size of the receiver only determines the fluctuation of the pressures in it. The smaller the receiver the greater the fluctuations.

It is obvious from what has already been shown that at the instant of cut-off of the condensing cylinder the pressure in it and also in the receiver, as well as the back pressure in the non-condensing cylinder, equals P_r.

At the instant of reaching the end of the stroke of the non-condensing cylinder it voids into the receiver its steam at a pressure eP_b.

If we wish this event to occur with as little disturbance as possible, we must make the pressure at that event equal in non-condensing cylinder and receiver.

The cranks being at right angles have certain definite positions with regard to each other at all times, and the position of one piston being fixed, that of the other can be deduced. If now we assume the cut-off e_1 of the condensing cylinder to be $\dfrac{1+k}{2}$, it will be coincident with the exhaust of the non-condensing cylinder, and the equation

$$P_r = \frac{eP_b}{e_1 R}$$

becomes $e_1 = \dfrac{1}{R}$; that is, $R = 2$, very nearly, for $e_1 = \frac{1}{2}$.

As the condensing cylinder completes its stroke under the pressure of expanded steam not connected with the receiver, its ratio can be somewhat greater, and consequently the ultimate expansions greater, without departing from the condition of equalized cylinder power.

The exit from the receiver to the condensing cylinder

being closed, the piston of the non-condensing cylinder now presses back the steam in that cylinder and the receiver until it has reached half-stroke; when the pressure is a maximum P_m, we can write the following equation:

$$eP_b\left(V_n + V_r\right) = P_m\left(\frac{1+k}{2}V_n + V_r\right).$$

In this equation we can fix P_m and determine V_r, or *vice versa.*

In general we can say that e_1 must be somewhat greater than $\frac{1}{R}$ in order to receive the exhaust steam from the non-condensing cylinder without forcing it back.

If we assume that the terminal pressure of the non-condensing cylinder must equal that in the receiver at the moment of opening communication, we can write the fol-

FIG. 42.

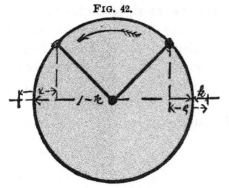

lowing equation, in which $x =$ the distance of the non-condensing piston:

$$P_r\{xV_n + V_r\} = eP_b(kV_n + V_r)$$

$$x = \frac{1+k}{2} - \sqrt{e_1(1+k) - e_1^2 - k}.$$

We have then

$$\frac{V_r}{V_n} = \frac{e_1Rk - \dfrac{1+k}{2} + \sqrt{e_1(1+k) - e_1^2 - k}}{1 - e_1R} = r.$$

At the instant of cut-off a quantity of steam is left in the receiver and non-condensing cylinder at a pressure P_r. As the cut-off e_1 is assumed earlier than half-stroke, we can then write the following equation:

$$P_r(x V_n + V_r) + e P_b V_n = P_m \left\{ V_r + \frac{1+k}{2} V_n + k V_c \right\}.$$

The two indeterminate quantities in this equation are P_m and V_r, either of which can be fixed and the other deduced. It will be observed that the clearance of the condensing cylinder is included, and therefore that P_m is the pressure at the instant the piston of the condensing cylinder is at the end of the stroke, its valve being assumed to have lead and a perfect vacuum to be obtained.

It will be observed that the presence of a receiver has the effect of reducing the power of the non-condensing cylinder when P_m is greater than $e P_b$, as it always should be. The greater the size of the receiver the less the increase of P_m; but this increase of P_m being made gradually, and the power taken from the non-condensing cylinder being restored to the condensing cylinder by reason of the increased pressure before cut-off occurs, it would not seem detrimental to economy to make receivers as small as possible.

The disadvantage of these high back pressures at mid-stroke in the non-condensing cylinder arises from the diminution of its power.

The only result of using a very large receiver, when cutting off steam in the condensing cylinder earlier than the point $\frac{1}{R}$, is to prevent the pressure in the receiver from rising much higher than the terminal pressure of the non-condensing cylinder.

A high back pressure at mid-stroke of the non-condensing cylinder also means a high initial pressure of the condensing cylinder, and, consequently, increased power in the condensing cylinder.

That is, a large receiver operates to prevent disproportion in the power of two cylinders when proportioned according to the criterion given in the preceding pages of this chapter, requiring also that the point $e_1 = \dfrac{1}{R}$ very nearly, but will be found not to be so economical of steam as a very small receiver, or none at all.

The clearances necessarily are regarded as receiver-space.

To sum up the discussion :

With cranks at right angles, we cannot cut-off later than one-half stroke in the condensing cylinder without a double admission to it, and consequent loss.

We cannot cut-off earlier than $\dfrac{1}{R}$ without pressing the steam from the receiver back into the non-condensing cylinder, because the steam will rise to a higher pressure in the receiver than the terminal pressure in the non-condensing cylinder. If this is done only to a small extent it may not prove a serious evil.

If we use a receiver of any considerable size, we must submit to a drop from the terminal pressure of the non-condensing cylinder to the pressure in the receiver, with the consequent loss, or make $e_1 = \dfrac{1}{R}$ very nearly.

If we do not use any receiver, or use only a very small one, we can effect a greater economy of steam ; but the power of the two cylinders cannot be equalized without pushing the ultimate expansion beyond all reasonable limits and diminishing the concentration of power so essential to all steam-engines, unless we make the point of cut-off of the condensing cylinder later than ½ stroke, and thus submit to a double admission or increase of pressure during expansion.

(92.) Condensation in Compounded Cylinders.—It

has been with the greatest difficulty that the writer has found data from engines proved steam-tight.

Indeed, it is safe to say that nearly every engine leaks badly, unless unusual precautions have been taken in fitting and direct experimental proof is given to the contrary.

The diagram of the example already solved will enable · us to follow clearly the steam-pressures and temperatures in the compounded cylinders.

The steam enters the non-condensing cylinder at a pressure of 100 pounds ($T_b = 327.5$), and is condensed upon walls having a temperature not lower than $230\frac{1}{2}°$, corresponding to a receiver pressure of 21.3 pounds. This condensation can be still further reduced by compression in the non-condensing cylinder, as shown.

When the steam has once been cut off in the non-condensing cylinder, all further demands upon the boiler cease for that stroke. The cylinders and receiver must satisfy their demands for condensation from what is in them already.

The final condensation in the condensing cylinder just before opening to exhaust is proportional to 204° ($12\frac{1}{2}$ pounds) minus 141.6° (3 pounds), and is thrown away through the exhaust, without appearing on the diagram.

Had a single cylinder been used of the same size as the condensing cylinder, the final condensation loss would prove the same; but the initial condensation for the same number of expansions and work would have been much greater, thus causing a larger volume of vaporous steam to disappear before expansion begins, and lessening the work (PV) done by a given weight of steam.

This is the explanation of the greater efficiency of compounded cylinders.

For any given conditions the comparison between single

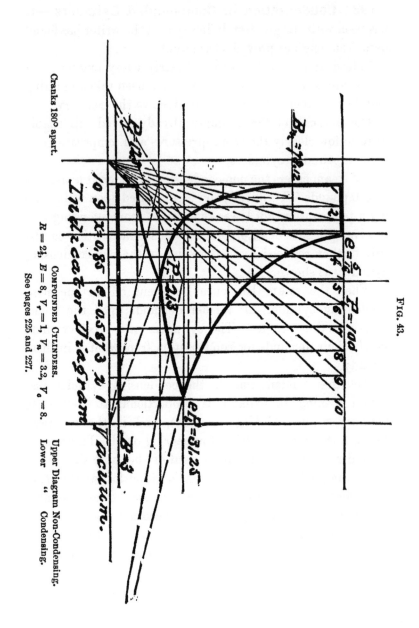

FIG. 43.

Cranks 180° apart.

COMPOUNDED CYLINDERS,

$R = 2\frac{1}{2}$, $E = 8$, $V_r = 1$, $V_n = 3.2$, $V_c = 8$.

See pages 225 and 227.

Upper Diagram Non-Condensing.
Lower " Condensing.

and compounded cylinders is made by means of formula (220 b) and analogous ones.

It is in improved methods of study and experiment that we must hope for advancement, rather than in the piling up of chaotic masses of more or less accurate, and almost always incomplete, experimental work, valuable only for commercial purposes. One series of experiments on a compound engine, carried out with care and skill, will give us a better basis than we have as yet for comparison with theoretical results, which we must use to guide us, knowing them not to be precise, but feeling them far more accurate than attempted generalizations from data which are in no wise comparable.

(93.) **Methods of Experimentation.**—In order to facilitate the great labor of arranging experiments upon boilers and steam-engines, the writer appends two codes, drawn up by him at the request of the Franklin Institute. These codes were adopted and used by the Institute during its International Electrical Exhibition of 1884, at Philadelphia.

(94.) **Code of the Proposed Quantitative Tests for the Evaporative Efficiency of Boilers at the International Electrical Exhibition, by the Franklin Institute, 1884.**

SPECIAL NOTICE.

Boilers may be exhibited and used at the International Electrical Exhibition, but will not have quantitative tests made of their efficiency unless formal application is made and the subjoined code accepted before July 15, 1884.

Competitive tests will not be made unless at the joint request of the parties desiring a competitive test, and after they have agreed to and subscribed to this code and fixed upon a rating for the points enumerated in Article 4.

The Committee of Judges reserve the right to limit the number of tests made, should time and opportunity not permit all the tests desired to be completed.

SECTION I.—PRELIMINARIES TO THE TESTS.

ARTICLE 1. *Capacity.*—The boilers entered may be of any capacity having an evaporative power not less than seven hundred and fifty pounds of water per hour.

Each boiler must be so drilled as to enable its whole internal capacity to be determined by being completely filled and emptied of water. *P*roper cocks, piping, etc., must be so placed as to enable this to be done readily.

ART. 2. *Pipes and Valves.*—Each exhibitor will furnish all the pipes and valves necessary to make connection with the main water- and steam-pipes in a proper manner, and subject to the orders of the superintendent. He will also make any alterations in water- and steam-pipes required for the tests, furnishing all tools, piping, cocks, and mechanical labor at his own cost.

ART. 3. *Space.*—Each exhibitor will be furnished with space at the regular rates established for the Exhibition, in which space he must build his foundations and boiler-setting, and make connection with the chimney-flue, if required, at his own cost, and subject to the approval of the superintendent.

ART. 4. *Specifications.*—Each exhibitor must furnish to the Chairman of the Committee of Judges on Steam-boilers such description and drawing, both of the boiler in position and of the details of the boiler, as will facilitate the labor of that Committee, together with his claims as to meritorious points for his exhibit.

The following points will have special consideration:

1. Economy of fuel.
2. Economy of material and labor of construction.

3. Evaporative power.   (Space occupied.)

4. Simplicity and accessibility of parts.

5. Durability of whole structure.

Exhibitors desiring a competitive test made must agree upon a rating for these points before it will be made.

Exhibitors must also file the following data:

Area of heating surface to the nearest hundredth of a foot.

Area of grate surface to the nearest hundredth of a foot.

Area of calorimeter to the nearest hundredth of a foot.

Area of chimney-flue to the nearest hundredth of a foot.

Height of chimney required.

Number of pounds of coal per square foot of grate to be burned per hour.

Should the calculations of the Committee of Judges differ in result from those of the exhibitor, he will be required to give all the details of his calculations, and an agreement must be reached before proceeding with the test.

SECTION II.—PREPARATIONS FOR THE TESTS.

ART. 5. *Coal.*—Anthracite coal will be used, and will be furnished free of charge, provided the steam made is used for the general purposes of the Exhibition.

The same quality and size of coal will be used in all the tests, unless special arrangements be made for another kind of fuel.

An analysis will be made of the coal used. The coal will be weighed to the boiler.

ART. 6. *Water.*—The water used will be taken from the city mains. The feed-water for the boilers will be weighed by means of scales and a large tank, and will be run into a smaller supplemental tank, from which it will be pumped into the boiler by means of a feed-pump actuated by steam from the boilers.

The temperature of the feed-water will be taken by means of a standard thermometer, in the supplemental tank.

ART. 7. *Pressure.*—The steam-pressure used shall not exceed ninety pounds per square inch by the gauge, unless by special arrangement with the Committee of Judges.

A standard gauge will be used, and also a standard thermometer immersed in a mercury pocket in the steam-space.

ART. 8. *Safety-Valve.*—The safety-valve will be set to blow off at ten pounds above the pressure fixed upon.

ART. 9. *Leaks.*—Within twenty-four hours preceding the test of a boiler, it must be subjected to hydraulic pressure ten pounds greater than its steam-pressure during the test, and proved to be perfectly tight.

ART. 10. *Attendants.*—The attendants in charge of the boiler tested must be approved by the party whose boiler is tested and by the Judges. All attendants are to be subject to the orders of the Judges during the progress of the test.

ART. 11. *Ashes.*—All ashes will be weighed on being withdrawn from the ash-pit, and must not be damped until weighed.

ART. 12. *Calorimeters.*—The calorimeters used will consist of a barrel, scale, and hand thermometer. Two calorimeters will be used, and simultaneous observations made at fifteen-minute intervals.

ART. 13. *Fires.*—The exhibitor shall be allowed one day previous to the test to clean boilers and grates.

The steam having reached the required pressure, the ash-pit shall be thoroughly cleaned and swept, and thereafter the fire maintained as nearly uniform as possible, the test closing with the same depth and intensity of fire as it opened.

This point is to be decided by the Judges, who may make allowance if it be clearly shown to have been impossible to maintain uniform fires.

If in the judgment of the Committee of Judges the firing

is inefficiently or improperly done, the test may be terminated at any time, and a repetition of the test refused.

Art. 14. *Pyrometer.*—The temperature of the gases of combustion immediately upon entering the chimney-flue shall be taken by means of a suitable pyrometer, read at fifteen-minute intervals, and close to the boiler.

Art. 15. *Manometer and Barometer.*—The vacuum in the chimney-flue shall be taken by means of a water manometer, read at fifteen-minute intervals. A barometer will be read simultaneously.

Art. 16. *Duration.*—Unless otherwise arranged, the tests will last ten hours.

Art. 17. *Economy and Efficiency of the Boiler.*—The level of the water in the boiler and the state of the fire must be kept as nearly constant as possible during the whole of the trial.

The weight of the water in the boiler for each one-quarter of an inch, on the glass water-gauge, will be carefully determined and recorded previous to the test, and proper correction for unavoidable changes of level made.

The weight of water fed to the boiler, subject to proper corrections, will be multiplied by its observed thermal value as steam. From this product the thermal units of heat brought in by the feed will be subtracted.

The remainder will be divided by nine hundred and sixty-six and seven-hundredths British thermal units, giving the number of pounds of water evaporated from and at 212° Fahrenheit.

This latter quantity will be divided by the weight of coal burned, less weight of dry ashes, giving the number of pounds of water evaporated per pound of combustible. This shall be taken as the measure of the efficiency of the boiler.

The nominal horse-power of the boiler will be deduced

by dividing the number of pounds of water evaporated from and at 212° Fahrenheit per hour by thirty.

The evaporative power of the boiler will be determined by dividing the nominal horse-power of the boiler by the number of cubic feet of space it occupies.

The space occupied by a boiler and its appurtenances will be regarded as the product of the square feet of floor-space occupied by its extreme height in feet.

(95.) Code of the Quantitative Tests proposed for the Steam-Engines at the International Electrical Exhibition, 1884, of the Franklin Institute, of the State of Pennsylvania.

SPECIAL NOTICE.

Parties exhibiting engines, who may desire quantitative tests made of them, must make formal application for such tests before July 15, 1884.

Engines can be exhibited, but will not be tested unless formal application and agreement to the code are completed within the specified time.

Parties desiring to have tests made of their engines can have them made by so signifying and by subscribing to and fulfilling the conditions of the code.

All tests will be quantitative, and will not be abridged, save by special agreement with the Judges.

Tests of regularity of speed will, however, be made independently of other measurements.

The Committee reserves the right to limit the number of engines tested, and to elect which engines shall be tested, if time will not permit complete tests of all.

Competitive tests will not be made save on the joint application of the two or more parties desiring them, who must agree on the rating of the various points of the en-

21

gines (see Article 9) previous to the tests, and subscribe in the code, agreeing to abide by the decision of the Judges without appeal.

SECTION I.—CONDITIONS OF EXHIBITION AND TEST.

ARTICLE 1. *Cylinders.*—The cylinders of the engines entered may be of any capacity and proportion of stroke to diameter.

ART. 2. *Indicator Connections.*—Each cylinder shall be drilled and tapped by the builder for indicator connections, by means of one-half inch pipe, in the usual manner, and to the satisfaction of the Judges. Pet drainage-cocks must be on the cylinder. The cross-head or other moving part must be drilled for the indicator cord attachment.

ART. 3. *Clearance.*—Each cylinder shall be drilled and plugged at both ends, so as to admit of being completely filled with water and emptied by means of a one-half inch pipe, in order to determine the clearance and the piston displacement of one stroke at each end. These data will be obtained both hot and cold.

ART. 4. *Valves.*—The steam- and exhaust-valves will be tested under full steam-pressure, ninety (90) pounds per square inch by the gauge, unless some other pressure has been agreed upon for the test.

ART. 5. *Piston Packing.*—The tightness of the piston packing will be determined by removing the back cylinder-heads and subjecting the piston to full boiler pressure on each centre.

ART. 6. *Fly-Wheel.*—Each maker is requested to use such diameter of band fly-wheel, or of pulley, as shall give a belt speed of 4000 feet per minute.

Should he require a different belt speed, he will specially note the same in communicating with the Exhibition Committee.

ART. 7. *Steam-Pipes.*—Each exhibitor will be required to furnish his own connection with the main steam-pipe, the main injection-pipe, and the main overflow-pipe or tanks.

ART. 8. *Space.*—Each exhibitor will be furnished with space at the regular rates established for the Exhibition, in which space he must build his foundations at his own cost, and subject to the approval of the superintendent.

ART. 9. *Specifications.*—Each exhibitor will communicate to the Chairman of the Committee of Judges on Steam-Engines such description and drawings of the engine exhibited as will facilitate the labors of that Committee, together with his claims as to the meritorious points for his exhibit.

The following points will have special consideration:

1. Economy of steam.
2. Regularity of speed.
3. Concentration of power.
4. Durability of construction.
5. Simplicity of design.
6. Excellence of proportions.
7. Finish of parts.

Each exhibitor must file the following data:

Diameter of the steam-cylinder to the nearest hundredth of an inch.

Diameter of the piston-rod to the nearest hundredth of an inch.

Diameter of the steam-pipe to the nearest hundredth of an inch.

Diameter of the exhaust-pipe to the nearest hundredth of an inch.

Diameter of the band, or fly-wheel, to the nearest hundredth of an inch.

Width of the face, or fly-wheel, to the nearest hundredth of an inch.

Weight of the fly-wheel in pounds.

Area of the steam-ports each to the nearest hundredth of an inch.

Area of the exhaust-ports each to the nearest hundredth of an inch.

Stroke of the engine to the nearest hundredth of an inch.

Indicated horse-power of the engine when working most economically.

Revolutions of the crank per minute.

Weight of the whole engine, exclusive only of the fly-wheel.

If a condenser is used and driven by the engine.

Diameter of the air-pumps to the nearest hundredth of an inch.

Diameter of the injection-pipe to the nearest hundredth of an inch.

Diameter of the overflow-pipe to the nearest hundredth of an inch.

Stroke of the air-pump piston to the nearest hundredth of an inch.

If an independent condenser is used that is not driven by the engine.

Diameter of the injection-pipe to the nearest hundredth of an inch.

Diameter of the overflow-pipe to the nearest hundredth of an inch.

Drawings of the condenser used, any other data peculiar to it, and a full description of it.

Section II.—Preparations for the Test.

Art. 10. *Steam.*—The steam for the tests will be furnished by the Exhibition boilers, and will come from boilers specially set apart for the purpose of the tests. It will be charged for at regular rate of three (3) cents per indicated horse power per hour. Steam, if desired, will be furnished to exhibitors one week before the tests are made.

No charge will be made for the services of attendants or experts, or the use of apparatus, unless in some extraordinary case, when the cost will be fixed by the superintendent of the Exhibition. No charge against the engine will be made for steam when its power is ordered by the superintendent for the other purposes of the Exhibition.

ART. 11. *Pressure.*—The steam-pressure used will be subject to the wish of the exhibitor, but shall not exceed ninety (90) pounds per square inch, by the gauge.

A special standard gauge will be used during the test, and subjected to careful tests before and after use.

ART. 12. *Safety-Valve.*—The safety-valve will be set to blow off at ten (10) pounds above the pressure fixed upon.

ART. 13. *Quality of the Steam.*—The thermal value, the temperature, and the pressure will be taken by means of scale calorimeters, thermometers, and standard gauges at the boiler, at the steam-chest, and at the exhaust, if the engine is non-condensing.

The thermometers, calorimeters, etc., will be furnished by the Exhibition, but the exhibitor must do such mechanical work, must furnish such piping, tools, and materials as are necessary to make the required attachments, at his own cost, and subject to the orders of the Committee of Judges.

ART. 14. *Temperature.*—The temperatures of injections and of hot-well will be taken with standard thermometers in the case of condensing engines.

ART. 15. *Water.*—The water used will be taken from the city mains.

The feed-water for the boilers will be weighed by means of scales and a large tank, and will be run into a smaller supplemental tank, from which it will be pumped into the

21 *

test boilers by means of a feed-pump, actuated by steam from other boilers.

The condensing water used will, in the case of condensing engines, be measured after leaving the hot-well in two carefully-gauged tanks, alternately filled and emptied, the temperature also being taken.

The known weight of steam used will be subtracted from the overflow.

The injection water will be weighed in large tanks, and its temperature taken.

The injection water will not be delivered under pressure.

ART. 16. *Speed of Engine.*—The number of revolutions of the engine will be taken by a continuous counter attached to the crank-shaft.

The variations in speed for one minute will be taken at each quarter of an hour by means of an electric chronograph connected with a standard clock beating seconds.

The variations in speed during one stroke will be taken by an acoustic chronograph at fifteen-minute intervals.

Special tests of speed alone, under varying loads, will be made if desired, and close attention will be had to this point in all cases.

ART. 17. *Barometric Measurements.*—A standard barometer and thermometer will be read at fifteen-minute intervals during the trial.

ART. 18. *Vacuum.*—The vacuum of condensing engines will be read by a gauge, carefully compared before and after the trials.

ART. 19. *Testing of Gauges, Indicators, etc.*—All of the gauges, indicators, and thermometers used shall be carefully tested before and after the trials, and the party whose engine is tested shall have the right to be present in person or by agent at these tests.

ART. 20. *Diagrams.*—The indicator diagrams will be

taken at fifteen- (15) minute intervals, and will be read for initial pressure, pressure at cut-off, terminal pressure, counter-pressure at mid-stroke, maximum compression pressure, mean effective pressure, point of cut-off, release of steam, exhaust-closure.

From the diagrams will be computed the indicated steam at the point of cut-off and at release, as also the actual steam from boilers per horse-power per hour.

ART. 21. *Load of the Engine.*—The Committee of Judges will test the engine at the load desired by the exhibitor of it, unless circumstances shall render it impossible to meet his wishes.

If the load does not exceed seventy-five (75) indicated horse-power, the net load will be measured by a transmitting dynamometer.

ART. 22. *Friction Diagrams.*—At the close of the regular trial, the engine will have its belt taken off, and be run for one hour for friction diagrams.

ART. 23. *Duration of the Trials.*—Unless otherwise arranged, the trials will last ten (10) hours.

ART. 24. *Economy and Efficiency of the Engine.*—No account will be taken of the coal burned; but the economy of the engine will be deduced from the actual steam used and water weighed to the boiler.

The trial will begin with the established pressure.

The level of the water in the boiler, and the pressure of the steam, will be kept as nearly constant as possible during the whole of the trial.

The whole weight of the water fed to the boiler, subject to proper deductions for waste, and to corrections for variation of level in the boiler, will be multiplied by its thermal value as steam at the steam-chest, and divided by the product of the indicated horse-power of the engine, and the number of hours of the test.

The resulting quotient will be used to divide twenty-five hundred and fifty-seven and sixty-nine one-hundredths (2557.69) British thermal units, giving the efficiency of the engine as compared with the mechanical equivalent of the heat furnished to it, and therefore its efficiency as a means of converting heat into work.

The net horse-power of the engine will be used for computation similarly to the indicated horse-power, and the result will be taken as the measure of the efficiency of the engine, both as a means of converting heat into work and as a machine for the transmission of power.

This latter shall be considered the true measure of the efficiency of the engine.

Lightning Source UK Ltd.
Milton Keynes UK
UKHW010955061118
331795UK00007B/277/P